ELECTRICITY
**Today's Technologies,
Tomorrow's Alternatives**

ELECTRICITY
Today's Technologies, Tomorrow's Alternatives

Revised Edition

EPRI Electric
Power
Research
Institute

Distributed by William Kaufmann, Inc.
95 First Street, Los Altos, California 94022

The Electric Power Research Institute (EPRI) was founded in 1972 by the nation's electric utilities to develop and manage a technology program for improving electric power production, distribution, and utilization.

EPRI is a nonprofit organization headquartered in Palo Alto, California. It is supported by voluntary contributions from both public and private utilities, and the results of its research are available to all.

The Institute's research is carried out in laboratories throughout the country—in private companies, in academic centers, in specialized research organizations, and in utilities themselves. The range of its work includes fossil and nuclear power plants, transmission and distribution technology, new forms of clean fuels, new energy sources such as solar and fusion, energy resources, and environmental problems connected with electric power technology.

Further information on EPRI and its research program is available from the Communications Division, EPRI, P.O. Box 10412, Palo Alto, California 94303.

Library of Congress Cataloging in Publication Data
Main entry under title:

Electricity, today's technologies, tomorrow's alternatives.

Bibliography: p.
Includes index.

1. Electric power.	2. Power resources.	I. Electric Power Research Institute
TK1001.E43 1982	621.31 82-2531	
ISBN 0-86576-042-X	AACR2	

Printed in the United States of America

CONTENTS

PREFACE

With publication of the first edition of this book in 1980, the Electric Power Research Institute, on behalf of the electric utility industry, took an important step toward public education in matters of electric energy technology. The immediate acceptance and continued demand for the book demonstrated the soundness of the project. This revised edition, available now with a teacher's guide for classroom use, furthers our educational efforts.

Of all the national and global issues that demand thoughtful public attention, none outranks energy. And electricity is the fastest growing energy form in the United States. Today, about one-third of our energy is converted to electricity before it is used. Looking ahead to the turn of the century, as much as half of our total energy will be produced in the power plants of the nation's electric utilities. That will not necessarily come about by choice. The need will, in large part, be forced on us by diminishing fossil fuel resources and world politics. Whether this momentum toward electricity will be sustained at a rational and adequate pace is an issue that demands full and immediate attention. And the technology of producing and delivering electricity cannot be separated from a discussion of how we get from today's forecasts to tomorrow's realities.

For as we try to make clear in the following pages, the old ways of producing electricity may either not be good enough in the future or not be available to us. Environmental goals, depleted domestic resources, and national security considerations are combining to create radically new technological ground rules for electric power producers.

As much as anything, then, this book provides a backdrop for public understanding of the issues as a prelude to development of new policies and initiatives. It tracks the relatively new role of electricity in our energy history, discusses old and new ways to produce it and related environmental issues, and closes with an agenda of technology-related issues that await decisions. Our purpose is served if *Electricity: Today's Technologies, Tomorrow's Alternatives* stimulates a measure of technical awareness and curiosity so that these decisions reflect the aspirations of an enlightened public.

In publishing this edition, we wish to acknowledge the expertise of technical writer/editor Suzanne Knott, who worked with EPRI technical and communications staff in revising material for the book and in writing the teacher's guide.

Ray Schuster
Director of Communications
Electric Power Research Institute

Chapter 1

ENERGY USE

Energy—the capacity to do work—comes in many forms and can be adapted to many uses. The energy that runs a car, for example, is different from the energy that runs a television set or a refrigerator. But one form of energy can be converted into another, and nature performs such conversions all the time—in the movement of a river current, in the growth of vegetation, in the evaporation of water by the sun. These natural processes and others that are under human control, such as electric power generation, are all based on changes in energy forms.

FORMS OF ENERGY

The forms of energy of most interest in electric power generation can be categorized roughly as chemical, heat, mechanical, potential, kinetic, nuclear, radiant, and electric (Figure 1-1).

Chemical Energy

Chemical energy is stored in a substance's chemical composition. Some compositions are better for storing energy than others. The fossil fuels—coal, oil, and natural gas, which are made up primarily of carbon and hydrogen—contain large amounts of chemical energy. This energy can be released by altering the substance's chemical composition—in the case of the fossil fuels, by burning them. A car's engine, for example, burns gasoline to release the chemical energy it holds.

Heat Energy

Certain chemical reactions, including those involved in the burning of fossil fuels, release heat energy. As a substance takes on heat energy, the molecules that compose it move faster and faster. Because heat energy transfers easily, one substance can be used to heat another. But heat energy is relatively hard to save and use at another time. For example, once the charcoal in your barbecue is burned to release heat, it is hard to recapture that heat for further use.

Mechanical Energy

Mechanical energy is energy expended by the application of force to an object, causing that object to move. A forklift uses mechanical energy to stack boxes in a warehouse. When you pull out weeds in the garden or push open a door, the mechanical energy of your muscles is doing the work. Similarly, you use mechanical energy when you lift a picture from the floor to hang it on the wall.

Potential Energy

When you hang the picture on the wall, the mechanical energy you expended in lifting it is retained by the picture in the form of potential energy. This is the energy stored in an object by virtue of its position. The picture has potential energy because of its high position relative to the floor. Water at the top of a dam also has potential energy because of its height.

Kinetic Energy

When water spills over the top of the dam, its potential energy is being converted to kinetic energy, the energy of an object in motion. Kinetic energy can be carried by any piece of matter—a molecule, a baseball, or a whole planet—as long as the matter is in motion.

Figure 1-1. Forms of Energy
Energy comes in many different forms. These are some of the most important for electric power generation. All of the nonelectric forms shown here are convertible into electric energy.

Nuclear Energy

Nuclear energy comes from the force of mutual repulsion among particles in an atom's nucleus. Normally, this force is counterbalanced by binding influences that hold the nucleus together. But during nuclear fission reactions, some particles break loose from the nucleus, carrying a great deal of energy with them in kinetic form. The same kind of particle release occurs in nuclear fusion, when the nuclei of two atoms join together. This kinetic energy can then be converted to heat energy and eventually to electric energy, although the conversion technology for harnessing the energy of fusion-type reactions is not yet fully developed (see Chapter 6).

Radiant Energy

The sun provides most of the radiant energy the earth receives. When they are heated, many substances give off light, a form of radiant energy. Wood, for example, gives off light when it is burned, and the light from a light bulb is produced by electrically heating a metallic filament in the bulb.

Electric Energy

Electric energy results from the movement of parts of atoms. The atom, which is the smallest unit of a chemical element, consists of a nucleus composed of protons and neutrons and surrounded by a number of orbiting electrons. (The hydrogen atom is an exception; its nucleus has only one proton and no neutrons at all.) The nature of an element is determined by the number and arrangement of its atom's components.

Protons are regarded as having a positive charge, and electrons, a negative charge. Like charges repel each other; unlike charges attract. The attractive force between oppositely charged protons and electrons holds the atom together. A normal atom is uncharged because it has the same number of protons and electrons and the charges of the protons and electrons are equal and opposite. However, electrons may be transferred from one atom to another, creating positive charges (more protons than electrons) or negative charges (more electrons than protons). An atom that has become charged by gaining or losing an electron is called an *ion*.

Just as a picture on the wall or water behind a dam has potential energy, a charged particle has potential electric energy by virtue of its position in an electric field. This potential electric energy is called *voltage*. Devices like generators and batteries create voltage by organizing ions into positive and negative accumulations. Like the potential energy of dammed water that becomes kinetic energy when the water flows over the dam, voltage can produce electric *current* when the two groups of charged particles are connected with a conductive material (most often metal wire). For example, in a battery the positive charges are gathered at one electrode, or terminal, and the negative charges at the other. When a conductor connects these electrodes, a circuit is formed (Figure 1-2). The negatively charged electrons, seeking their opposites, migrate from the negative electrode (the cathode) toward the positive one (the anode). The flow of electrons constitutes electricity.

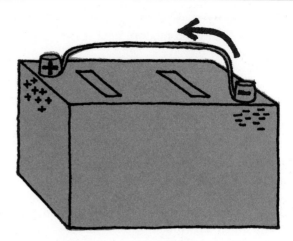

Figure 1-2. The Formation of Electric Energy
Voltage is potential electric energy. Batteries create voltage by accumulating negative and positive ions into groups. When the two groups are connected by a conductive material such as metal wire, the charged electrons flow through the wire, creating electric current.

ENERGY USE THROUGH THE AGES
Human beings have used some of these energy forms since before recorded history. Others have been harnessed only recently.

Preindustrial Life
People have always been able to use their bodies' mechanical energy. The fuel for that energy is the conversion of the chemical energy in food by digestion; the sun imparts radiant energy to plants through the process of photosynthesis. To obtain their food, the first people followed the great herds of beasts and gathered wild vegetation. These people learned very early to convert the chemical energy of fuels, such as wood, to heat and light. Primitive people used these energy forms for simple survival—for warmth and for cooking.

When animals were domesticated, their strength and mobility were used to do work. They were harnessed to crude plows to till the soil. Nomads settled into communities and began cultivating their own food. Civilization was born and flourished because the food supply could be managed. Surplus meant that the community could support individuals who did not produce food. Political, social, and economic structures developed that affected all aspects of life. People learned to build dams to harness the kinetic energy of water for crushing grain and to design sails to catch the wind and power ships.

Although individuals in every part of the world continued to invent devices that saved their own energy, the effects of such tools spread slowly (Figure 1-3). Life was hard work. Planting and harvesting were back-breaking. Manufacturing processes were carried out by hand or by crude machines operated by hand.

However slowly, the effects did spread, carried by continuous social, political, and economic changes. And with them came new knowledge and new tools. As knowledge and the use of tools increased, so did their effects. The *rate* of change increased. Knowledge fed on itself; discovery followed discovery—faster and faster. By the eighteenth century knowledge and discovery produced a device more powerful than humans or animals. It mechanized the means of production and industrialized the western world.

The Industrial Revolution
The steam engine converted the chemical energy of a fuel (wood, coal, and, later, gas and oil) into the heat energy of steam and then into the mechanical energy of moving pistons, which could drive all kinds of machinery. These diverse and powerful machines made possible the division of labor and the factory system, particularly in the young United States, where resources were abundant and labor scarce. Factories could produce more goods more inexpensively than individual craftsmen could. Markets expanded and more commodities became available. The standard of living rose. People were

freed from dawn-to-dusk labor to pursue educational and recreational activities. Colleges and universities proliferated; games and modern team sports became popular.

The mechanization of production sparked interest in applied technology in all aspects of living. In the 1800s the steam engine was applied to transportation. Steamships and locomotives made travel faster and more reliable. New fuel sources were discovered. Gas and oil drove the steam engine. A new type of motive power—the internal combustion engine—was developed that was to transform transportation years later. Volta built the first battery, and Faraday developed the dynamo—an effective means of electricity generation. Communication was enhanced by the advent of the telegraph and, later, the telephone. The day was extended by means of abundant artificial light made possible by kerosene and natural gas.

This zeal for technological innovation manifested itself in the pursuit of incandescent light—light produced by the steady burning of a filament heated white-hot by an electric current. Interest in electric energy had begun some 200 years earlier, when electricity was first produced in a crude electrostatic generator. There followed years of investigation into the nature of this spectacular phenomenon and ways to produce it. By the nineteenth century, Davy had demonstrated that incandescence was possible. But continuous light from an electric current remained elusive, although many people were trying to produce it. In 1879 the goal was achieved. Edison and his colleagues perfected the incandescent light, introducing a new age.

ENERGY USE IN MODERN SOCIETY

Effects of Electricity

The age ushered in by the incandescent light was powered by electricity. Although the incandescent light is its most visible aspect, electric power affects modern life in every way. This convenient energy form, available to each of us at the flip of a switch, is used in almost countless ways. Electric motors and other electric-powered devices have replaced most steam-powered factory equipment. Almost any item we use in our daily lives is available in its present form because of electricity: clothing, food, building materials, entertainment. Electric power helps make them all possible.

We even organize our lives around electricity. The very structure and use of buildings takes electricity for granted. Electric power provides heat and air conditioning; we move between floors by electrically driven elevators. Many forms of transportation, such as subways, rely on electric power. Modern air travel would be

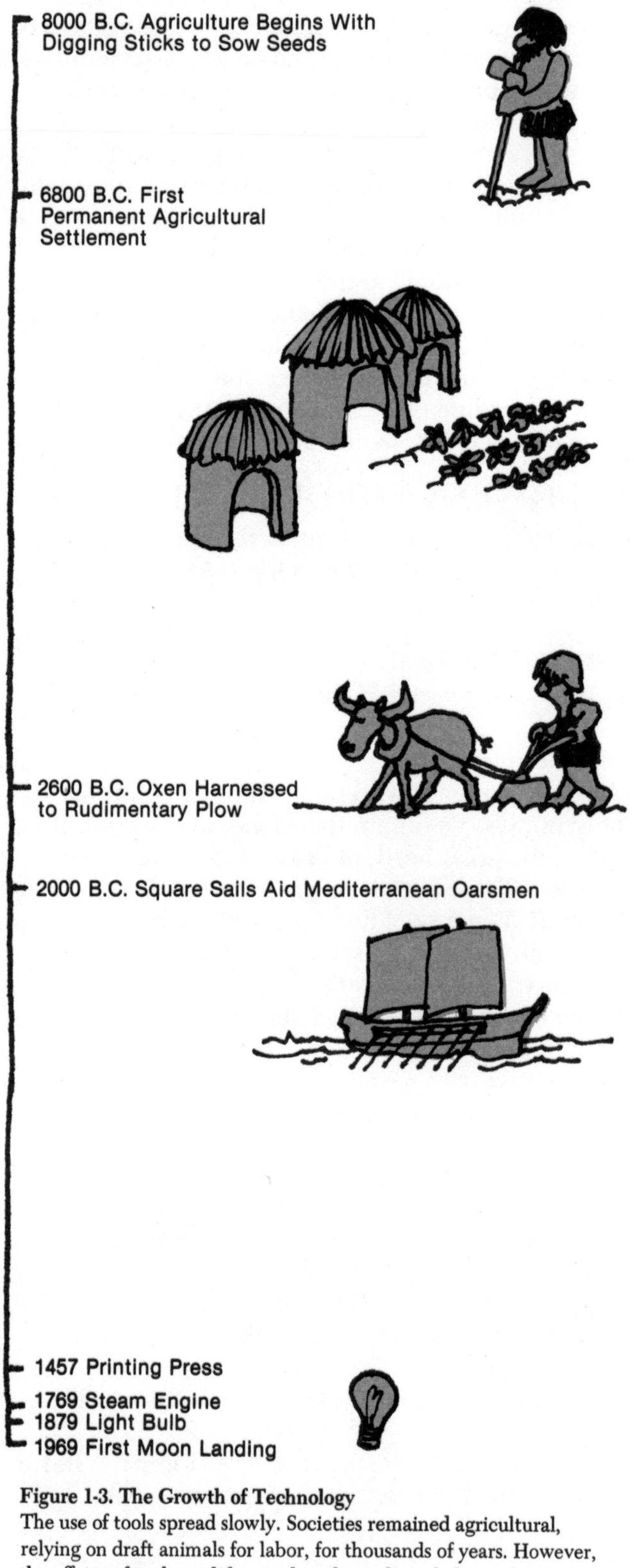

8000 B.C. Agriculture Begins With Digging Sticks to Sow Seeds

6800 B.C. First Permanent Agricultural Settlement

2600 B.C. Oxen Harnessed to Rudimentary Plow

2000 B.C. Square Sails Aid Mediterranean Oarsmen

1457 Printing Press
1769 Steam Engine
1879 Light Bulb
1969 First Moon Landing

Figure 1-3. The Growth of Technology
The use of tools spread slowly. Societies remained agricultural, relying on draft animals for labor, for thousands of years. However, the effects of tool use did spread, and new knowledge and new tools resulted. In modern times the rate of discovery and invention is astounding.

impossible without runway lights and communication with control towers. Electricity runs computers for business, science, and home use. It provides entertainment. It helps in the diagnosis and treatment of disease and provides the means for miraculous surgical feats. Electricity makes possible instantaneous global communication. It allows us to explore the vastness of interplanetary space, as well as the infinitesimal world of atomic particles.

Modern Energy Use

In addition to electricity, other energy forms have been made available to us by scientific investigation within the last century. And technology, largely affected by electric power, has allowed us to put these forms to work. Discoveries in radiant energy have resulted in many inventions using electromagnetic waves. Radio, television, and microwave ovens have become common household items. X rays and infrared light are used extensively in medicine, scientific research, and industry.

The work of twentieth-century scientists has enabled us to split the atom and capture and use the nuclear energy released by this fission process.

The way we use energy forms has changed immeasurably since our nomadic ancestors followed animal migrations. Not only do we use forms that were available then in a different way, we also have access to forms undreamed of by earlier people. We use the chemical energy of fossil fuels for heat, transportation, and manufacturing. Our dependence on electricity requires that we generate it by converting the chemical energy of fossil fuels into the heat energy of steam into the mechanical energy of turbine generators. Or we convert the heat energy of nuclear fission. Or we convert the kinetic energy of falling water at dams. Modern energy use is inextricably bound to technology; neither one would be possible without the other.

Chapter 2

ENERGY DEMAND

Our society depends on energy. In fact, we need a lot of energy if we are to live comfortably and do all our tasks: produce our food, construct our buildings, maintain a system of transportation, and manufacture all the goods we use in our daily lives. To ensure that we have enough energy to do all these things, we have to understand how much energy we use and how much we are likely to need in the future. We have to understand energy demand.

DEMAND MEASUREMENT

Energy Demand

Energy demand—the amount of energy consumed—is measured by a unit called the *British thermal unit*, or Btu. This measurement represents the amount of heat energy it takes to raise the temperature of 1 pound of water by 1 degree Fahrenheit. One Btu may seem like a small amount of energy, but the number of Btu it takes to produce all the goods and services we use is very large indeed.

In fact, the number of Btu consumed is so large that an informal measure called a *quad* is used to talk about energy consumption. Quad is short for one quadrillion Btu—1,000,000,000,000,000 (or 10^{15}) Btu. Every quad is the equivalent of about 172 million barrels of oil, 980 billion cubic feet of natural gas, or 40 million tons of coal. It represents roughly the amount of energy consumed in a large metropolitan area such as Boston, San Francisco, or Pittsburgh in one year.

Electricity Demand

Electricity demand is measured by the *watt*, which is a unit that indicates the rate at which electric power is generated or used in an instant. Because the watt is a small unit, kilowatts (thousands of watts) are often used. Very large amounts of electricity are measured in megawatts (millions of watts) or gigawatts (billions of watts).

If we turn on a 1000-watt hair dryer, we use 1000 watts of electricity at that moment. But wattage does not tell us how much electricity we use during the time it takes us to dry our hair. The watthour has been devised to measure the amount of electricity used over time. One watthour is the use of 1 watt of electricity for 1 hour. Drying our hair for 1 hour with a 1000-watt hair dryer would use 1000 watthours, or 1 kilowatthour (kWh), of electricity.

Present Demand

Total energy demand is the demand for all energy we use for all purposes—for industry, transportation, and residential and commercial use, as well as for electricity generation. How much energy do we use as a nation? Figure 2-1 shows us. Industry demands slightly more than one-quarter of all energy—21 quads, or 27%. It takes tremendous amounts of energy to run all the country's machinery and produce the myriad products we use. Transportation uses 19 quads, or 24%, almost as much as industry. It's easy to see why. There are millions of cars in the United States, as well as large public transit systems in all the major urban areas. Hundreds of jets take off from U.S. airports every day. All those vehicles consume lots of energy. Commercial and residential demand accounts for about 12 quads, or 16% of the total. This figure includes all the energy required for heating homes, schools, and offices. The greatest amount

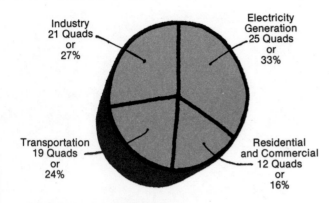

Figure 2-1. Total U.S. Energy Use
Industry and transportation demand more total energy than do homes and offices. Of all transportation use, the private car consumes the most energy. Our biggest total energy demand is for electricity generation.

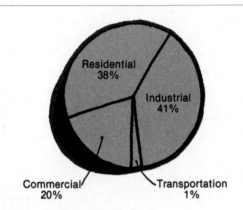

Figure 2-2. U.S. Electricity Use
Most of our electricity goes to residential and commercial use. Industry's share will probably increase as new manufacturing processes that use electricity are developed. Transportation, which relies heavily on oil, uses relatively little electricity.

of energy goes to generating electricity. It takes about 25 quads of energy, or 33% of total demand, to generate all the electricity we use in the United States.

We consume enormous amounts of electricity: about 2.4 trillion kWh each year. Figure 2-2 shows how we use that electricity. There are 84 million houses in the United States, and each one uses about 11,000 kWh each year. So our home use is 910 billion kWh, or 38% of all electricity used. In addition to the other energy sources it uses, industry consumes 980 billion kWh of electricity to run its machinery. Industry's share is 41%. Transportation, which uses more oil and less electricity, uses 10 billion kWh, or less than 1%, most of which goes to mass transit systems such as subways.

ENERGY AND ECONOMY

Gross National Product

As we can see, everything we do in our daily lives depends on energy use, and our energy needs are closely tied to the country's economic activity. The state of the economy is reflected by the gross national product (GNP), which is the total value of all goods and services produced in a nation in one year. When we consider how much energy is required to produce all our goods and services, the relationship between energy demand and GNP becomes obvious. In fact, we can see it clearly in a chart (Figure 2-3). Energy demand increased steadily during the early part of the 1970s. The economy also grew at a fairly even rate. After 1973, however, demand and GNP fluctuated. A contributing reason may have been an increase in the price of oil.

The Role of Price

Price affects demand—and the economy. How? Economists define demand as the quantity of an item that people are willing to buy at a certain price. If the price drops, demand increases because more people are able to afford a product and more of the item is sold. If the price increases, demand decreases; fewer people can afford the product. In 1973 the oil-exporting countries of the world began raising oil prices sharply. There was a corresponding drop in U.S. energy use, particularly for transportation, as the nation cut back its oil use. As a result, U.S. economic growth slowed, which meant a reduction in goods and services (Figure 2-3).

Changes in the price of a resource affect not only the economy as a whole but also the demand for other resources. If, for example, we could produce the same amount of a product using equal amounts of oil or coal and the oil were cheaper, we would certainly use the oil. But if the prices were about the same and coal were easier to obtain, we might use coal, knowing we could always get as much as we needed to manufacture our product and stay in business. The oil price increases caused consumers to look for other energy sources that had traditionally been more expensive than oil.

While total energy demand and the economy declined, electricity use continued to increase (Figure 2-4). Electricity has been an inexpensive power source, partly because fuel sources to generate it have been inexpensive until the past few years and partly because utilities are able to generate large quantities of electricity in central stations, making the cost of a kilowatthour as small as a few cents. Further, electricity is a high-quality, convenient, and clean energy form, and the number of ways it can be applied continues to grow.

How will rising prices and the quest for substitute energy sources affect future demand? Will energy demand increase? What about electricity use? The answers to these questions are important for several reasons. Energy demand is closely tied to the economy; economic growth depends on energy. Unlike other products, which are manufactured and placed in a warehouse until people want them, electricity must be made the moment it is to be used. Since utilities cannot store electricity, they must have very accurate projections of how much electricity people may want so that sufficient generating capacity is available. (Some storage is possible in batteries and pumped storage reservoirs, but the amount is relatively small compared with the amount required at any given moment.) Because it takes a great deal of time and money to add new generating capacity, energy requirements must be anticipated years in advance.

DEMAND FORECASTS

Making predictions is difficult because there are no facts about the future. Predictions about future energy demand are especially difficult. So many factors—social, political, and economic—affect that demand. Natural forces also greatly influence energy use. A particularly hard, cold winter; a drought; a long sweltering summer: each will change energy use patterns. All these factors—natural and cultural—must be taken into account in making energy demand forecasts.

Future Energy Demand

To forecast demand, we must make certain assumptions about all these factors for a particular point in time. The forecasts made by experts vary widely because indi-

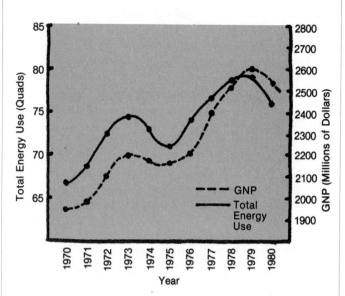

Figure 2-3. Energy Demand and GNP
Energy demand is closely related to economic activity. Both total energy demand and the economy grew in the early 1970s. Large price hikes for oil—one of our main energy sources—caused consumers to reduce demand, and the reduction was reflected in a slowing economy.

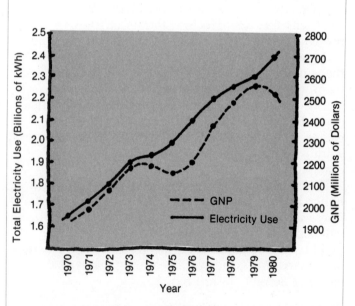

Figure 2-4. Electricity Demand and GNP
Electricity demand rose consistently throughout the 1970s, in spite of dips in both total energy demand and GNP. Electricity's versatility and relatively low cost account for its extensive use as a power source.

viduals select different assumptions about which factors are most linked with energy demand and how each of these factors will behave.

Experts also vary the assumptions they select and make up several accounts of what they think will happen. These accounts are called scenarios. For example, let us assume that the economy will grow at a rate of 2.5% per year between now and the end of the century. In this scenario, we may use 86–88 quads of total energy each year by 1990 and 101–106 quads by the year 2000 (Figure 2-5). If the economy grows at a faster rate—say, 2.9% each year—we will be using more energy. In this scenario of higher economic growth, we may use 92–94 quads in a year by 1990 and 114–121 quads by 2000.

Future Electricity Demand

Much of that energy will be used to generate electricity. As the population continues to increase, electricity demand will increase. Where people live will also affect electricity use. The population has begun to shift from the Northeast and Midwest to the West, Southeast, and Southwest, and this trend is expected to continue. The Northeast and Midwest are less electricity-intensive. Residential heating in these areas, for example, is fueled primarily by oil and gas. The growing population in the West and South, on the other hand, creates a growing demand for air conditioning provided by electricity.

Changes in the way we produce our goods and services will also account for some of the expected growth in electricity use. New processes for industry will be powered by electricity. The number of light manufacturing and service industries is expected to increase, and these industries are electricity-intensive. Technological innovations will allow many industries, businesses, and households to use more processes and applications that depend on electricity. Home computers, television games, robots for manufacturing processes, and advanced worldwide communications systems are just a few examples.

How much electricity will we use? We may need 3–4 billion kWh in 1990 and 3.5–6.4 billion kWh by 2000. These are vast amounts of electricity, and our well-being will depend on our ability to meet the demand. The problem of meeting future demand has focused attention on conservation as one way of ensuring that we have enough electricity for the future.

CONSERVATION

How we should use our energy to achieve the most benefit has become a very important question. The increasing cost of energy production means that con-

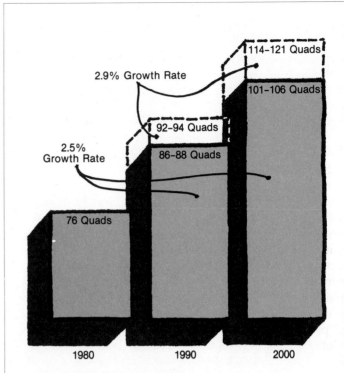

Figure 2-5. Future Energy Demand
Many forecasters see U.S. energy use increasing substantially over the next two decades. Different projections about future energy use reflect different assumptions about how we will live, how efficiently we will use energy, and how fast the economy will grow.

servation must play a part in energy planning. But conservation cannot simply be a reduction in energy use. As we can see from our experience in the 1970s, cutting back energy use hampers economic growth, and the result is fewer goods and services, fewer jobs, and a lower standard of living.

Conservation and GNP

The aim of conservation should be to increase the *efficiency* of our economic activity—that is, to get more economic growth (a larger GNP) from *all* the elements we put into production of goods and services: capital, labor, energy, and materials. We can do this by devising technological changes and substitutions that increase the value of our production output in relation to what we put into the production process.

In the last 50 years we have become more efficient. The ratio of total energy demand to GNP has decreased (Figure 2-5), and continued improvements in energy efficiency could reduce the amount of energy we need for economic growth by as much as 26%. Electricity will play an important role in this increased efficiency. Technology has pushed many industries, businesses, and households toward processes and applications that depend on electricity. Even though some of these technologies may require more electricity, they may reduce reliance on other costly energy sources or on the other elements of production, and so they may spur overall economic growth.

Conservation and Energy Demand

U.S. industry provides some examples of how technology can increase production efficiency. Steel industry furnaces are designed to operate on high-grade iron ore with an iron content of 50% or more. Such quality resources have been dwindling for some time, however, and available lower-grade ores contain less than 30% iron. The industry was compelled to use costly imported ore or convert lower-quality ores, which required more material, produced large amounts of waste, and required greater amounts of energy. Now, with the development of processing plants at the mines, the ore can be collected and shaped into pellets of greater iron concentra-

tions. The processing plants require electrically driven pumps, balling machines, magnets, and other equipment. However, the process reduces the need for high-grade material and the demand for imported ore, as well as lowers the blast furnace fuel requirements and reduces energy consumption at the blast furnace by nearly 20%.

Until recently, the outside coating and decoration on common cans was dried by curing in gas-fired ovens. Most cans were made in three pieces from previously coated material. New two-piece can systems cannot be made from previously coated material, but special inks and coatings that can be cured by ultraviolet and infrared light have allowed manufacturers to coat and decorate these new cans. The new technique consumes greater amounts of electricity but substantially reduces consumption of other energy sources.

Although we can use our energy more productively, we will still require huge amounts in the future. To meet these energy needs and to ensure that the economy will continue to grow, we will need adequate and reliable supplies. Let's look now at energy supply, at what elements affect our use of resources, and at what kinds of supplies we may be using in the future.

Chapter 3

ENERGY SUPPLY

Fueling today's energy demand requires vast supplies of fuel. To ensure our ability to meet future demand, we must ensure adequate and reliable supply sources. But first we need an understanding of what elements affect resource use, what resources are available to us, and how technology can help us extend our resources for the future.

RESOURCE USE

Our energy supply comes from many different sources. The fossil fuels (oil, natural gas, and coal) and uranium come from the ground. Renewable resources, such as the sun, wind, and rivers, are continuously or cyclically renewed by nature. Electricity is a manufactured form of energy, and its generation is fueled by the other, natural sources.

The amount of energy we obtain from all these sources depends on several considerations: availability, price, technology, and environmental effects. They govern the ways we supply all our energy needs, including the way we fuel electricity generation. Oil provides an excellent example of how these elements influence resource use because oil supply has changed dramatically in recent years and has affected all of us.

Availability

In the early part of this century large accessible oil deposits in Texas and Oklahoma could be tapped with ease—and they were. The demand for oil grew rapidly. Its liquid form made it easy to handle, and the burgeoning automobile industry provided a ready market. But as energy demand grew in the 1950s and 1960s, the yield from those rich fields was not sufficient to meet it. Other deposits that are smaller, lower in grade, harder to find, and harder to tap were not developed because quality oil was available elsewhere in the world. We began to import crude oil from other countries.

By the early 1970s we were using huge amounts of oil—about 17 million barrels a day—and about one-third of it was imported. In 1973 the oil exporting nations temporarily embargoed oil; that is, they stopped shipping oil overseas to countries such as the United States. The embargo seriously jeopardized our ability to meet energy demand. Subsequent events in the politically troubled Middle East—where we get a lot of our imported oil—have made it clear that oil supplied by that part of the world may not always be available. Availability is a key element in resource use; if we cannot obtain a resource, we cannot use it. But if one resource is not available, others will be. The role of price becomes very important.

Price

Economists define supply in terms of price, and they commonly argue that resources will always be available —if the price in the marketplace is high enough to cover the cost of recovering them. But of course there are practical limits to what consumers can pay. Before the mid-1970s, oil was abundant and inexpensive compared with other energy sources. When the oil-producing nations resumed trading after the embargo, they began raising the price of oil. Since that time, our oil import bill has undergone a tenfold increase. Higher foreign oil prices made more expensive sources—such as the harder-to-get oil deposits in the United States—more attractive. Price provides the incentive to find and develop resources that may be far away from where they will be used, in harsh climates, very deep under ground, under water, or simply of lesser quality. Price measures the extent to which we have exhausted the cheaper and easier-to-find resources. It determines what resources we can afford to develop through improved technology.

Technology

Technology can make accessible those resources that might otherwise go unused. For example, the United States has vast deposits of oil shale in the West. The Green River formation, a geologic feature encompassing parts of Colorado, Wyoming, and Utah, has been estimated to contain the equivalent of about 1.8 trillion barrels of oil. Extracting oil from rock is naturally far more complicated than pumping liquid oil out of the ground.

The technology is available, however. In one process, the rock is mined, crushed, and retorted—that is, heated to about 900°F in a special vessel. At that temperature the embedded oil decomposes into gas and liquids, which can then be refined conventionally. This particular technology is not very complicated. The problem is the massive size of the operation: about two tons of rock must be mined to produce one barrel of oil. Although technology can help us obtain less accessible resources, we must consider possible damaging effects on the environment.

Environmental Effects

The effect on the environment of the development and use of a particular resource may determine the extent of its use in energy supply. Oil shale mining and crushing processes may release dust and particles that violate existing air quality laws. In addition, the mined rock itself undergoes a "popcorn effect" during processing, which means that it expands to about 150% of its original volume, leaving that much more rock to dispose of after the oil is extracted. How to dispose of the virtual mountains of leftover shale is no small problem. This technology also uses large amounts of water, a resource that is already scarce in the West. These problems, together with fears about the impact of oil shale processing on local plant and animal life, have proved to be major obstacles to our use of domestic oil shale resources.

Processes for extracting oil from shale while it is still underground promise some relief from these environmental problems. Several developers propose to blast out

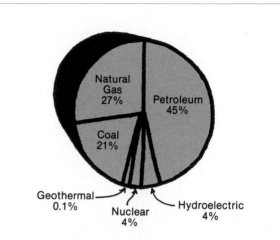

Figure 3-1. U.S. Energy Supply Sources
We use more petroleum than any other supply source. Natural gas and coal are also important sources. These three—the fossil fuels—supply us with a whopping 93% of total energy.

giant underground caverns, then fire them to release the oil from the shattered rock. This task will not be easy. A big question is how much of the oil present can be recovered in this way. However, keeping the process underground helps sidestep air quality problems as well as the task of trying to revegetate heaps of spent shale on the land's surface. A potential problem with such processes is contamination of underground water supplies.

Availability, price, technology, and environmental effects determine whether we can use certain resources to supply our energy needs, as well as how much of these resources we can use. What resources are available to us and how do we use them?

SUPPLY SOURCES

In 1980 the United States consumed 76 quads of energy. Figure 3-1 shows what fuel sources provided that energy. We used more petroleum than any other source—34 quads, or 45%, almost as much oil as all other energy sources combined. Natural gas and coal are the second and third most important fuel sources; we used 20 quads of natural gas, which was about 27% of the total, and 15 quads of coal, about 21%. Nuclear and hydroelectric power each provided about 3 quads, or 4% of the total. Geothermal use was 0.1 quad. How much of our energy will each of these resources supply in the future, and how will availability, price, technology, and environmental effects determine their use?

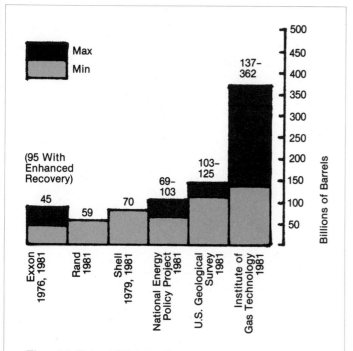

Figure 3-2. Estimated U.S. Petroleum Resources
Estimates of undiscovered or unproven petroleum resources cover a broad range. Future oil production will come from these resources plus known reserves. Note that estimates with different dates should not be compared directly, since any new oil finds that may have occurred between the two dates would transfer undiscovered resources into the category of known reserves.

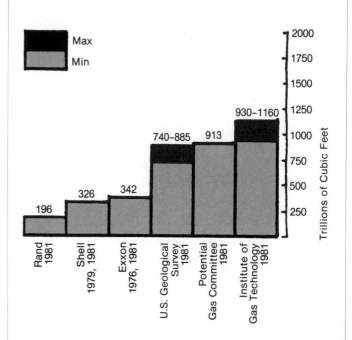

Figure 3-3. Estimated U.S. Natural Gas Resources
As with those of oil, estimates of undiscovered or unproven natural gas differ widely. Even if we could determine the physical extent of these resources accurately, recovering them would remain a problem. Price, technology, and environmental constraints help determine how much of our resource base we will actually be able to recover and use.

Oil

Higher oil prices have allowed oil producers in the United States to drill deeper wells for harder-to-get and more expensive oil. The number of deep wells (below 15,000 feet) drilled in the United States has been increasing since 1977. Total U.S. drilling has now reached 80,000 new wells each year.

How large are U.S. oil reserves? Estimates range from 45 billion barrels to 137 billion barrels (Figure 3-2). No one is certain just how much of this oil we may actually recover. We will have to use other resources in place of oil, or we will have to develop technology that will allow us to obtain oil from shale or liquid fuels from coal economically.

Natural Gas

Higher prices for natural gas have also brought deeper drilling and exploration in new areas. No one knows how higher prices will improve supply, although deep gas, gas in tight formations, and gas from frontier areas will begin to play an important role. Estimates of gas reserves range from 400 trillion cubic feet to 1200 trillion cubic feet (Figure 3-3).

Coal

Coal is our most abundant fossil fuel resource (Figure 3-4), and its share of energy production is expected to increase as that of oil declines. The federal government recently estimated that we have coal reserves of 475 billion tons, one-third of which can be mined from the land's surface. The portion that can actually be recovered and marketed is uncertain because mining

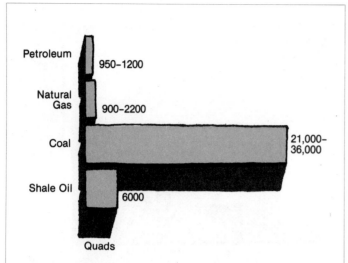

Figure 3-4. U.S. Coal Reserves and Resources Compared With Other Fossil Fuels
Coal resources in the United States far exceed resources of other fossil fuels. That is why energy planners are encouraging the use of coal, wherever possible, instead of oil or gas. Greater reliance on coal-fired electricity generation and the possibility of synthetic fuels from coal make this fuel an important energy supplier.

conditions and the quality of coal vary considerably from region to region.

One problem with coal is that its use affects the environment in several ways. Surface mines result in huge open pits, and the sites must be restored once the mines have been played out. In addition, burning coal produces various gases (sulfur dioxide, nitrogen oxides, and carbon dioxide) and particulates, which are emitted into the atmosphere from the stacks. Coal-cleaning processes increase the price of the fuel, and equipment to remove the gases and particulates during combustion increase the cost of using it.

Future coal use will also depend on technology to deal with the structure of coal itself. For many of our energy uses, particularly for vehicles, we require a liquid fuel. So that oil will be available for transportation, many consumers, such as electric utilities, are being asked to use coal instead whenever possible. But utilities sometimes have to use liquid fuels to run generators to meet peak loads that last only a few hours each year. So researchers are working on ways to process coal to produce liquid fuels that could be used in utility boilers and perhaps eventually in transportation as well.

Nuclear

Commercial nuclear power in the United States is facing a period of stringent evaluation. Licensing delays, questions about radioactive waste disposal, and concern about nuclear power plant safety are major obstacles to the growth of this energy source. However, nuclear power is essential to our future energy supply.

One important factor is that uranium, which fuels the fission reaction, is a domestic fuel. U.S. uranium resources are substantial—about 2 million tons. Further, if we use the fast breeder (see Chapter 6), our nuclear fuel supplies could be extended about 60 times. Nuclear power is also one of the least expensive ways to generate electricity.

Just how much of our energy supply will be contributed by nuclear power is uncertain. Experts recognize the seriousness of the obstacles to nuclear power expansion, but they also recognize that serious electricity shortages could result if nuclear power is excluded from our energy supply.

Renewable Resources

Renewable resources—sun, wind, rivers, oceans, heat from within the earth, organic wastes produced by natural and technological processes—are power resources that are constantly replenished. Unlike other sources we use, they seem inexhaustible, and that quality has kindled great interest in harnessing them. Technology and cost are both obstacles to their use.

Only two—hydroelectric and dry steam geothermal—currently contribute to our energy supply. Hydroelectric power generation has been in use for almost a century. Recent advances in technology and construction methods, together with the rising prices of fossil fuels, have made small hydroelectric projects seem practical. Federal incentives and streamlined licensing are also encouraging hydroelectric power development. Although almost all sites that can accommodate large dams have been developed, small-capacity sites are being explored.

Geothermal dry steam is vented from beneath the earth's surface. Although such a resource can be tapped with relative ease right at the surface, there are only three known steam fields in the United States. Two are in Yellowstone and Lassen National Parks, respectively, and the third, in California, is the only one available for commercial development.

The technologies for other renewable resources (biomass conversion; wind power generation; solar-thermal conversion; photovoltaic conversion; exploitation of geothermal hot water, geopressured zones, hot rock, and magma; and fusion) are in various stages of

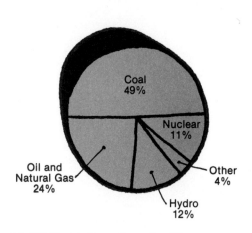

Figure 3-5. U.S. Electric Power Generation Mix
Present sources of electric power show coal in the lead, with sizable shares still supplied by oil and gas. Nuclear power provides about 11% of our electricity, and hydropower a significant slice of renewables. Other renewable resources, such as dry steam geothermal, still make only a token contribution.

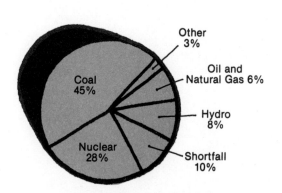

Figure 3-6. Projected Electric Power Generation Mix
EPRI projections show electric power generation in the year 2000 relying more heavily than ever on coal. Increased nuclear capacity is also expected to help replace oil and gas. Hydro's share will have shrunk because of siting limitations, but other renewables will contribute more. The shortfall (demand that cannot be supplied) could be supplied with more nuclear, coal, or oil generation; more conservation; or reduced economic activity.

development. Whether the technologies can be developed and whether these sources will be economical is uncertain.

Generation Mix

Electricity will supply more and more of our energy. In addition to its uniqueness and versatility, electricity can be—and is—generated from a number of different fuel sources. The proportions of these sources that utilities use to produce the nation's electricity determine the *generation mix,* and this mix is expected to change in the next 20 years as availability, price, technology, and environmental effects determine fuel use.

Figure 3-5 shows the U.S. electric power generation mix in 1980, and Figure 3-6 shows the projected electric power generation mix for the year 2000. By 2000, the use of oil and gas, including liquid fuels derived from coal, may be sharply reduced, from 26% of total operation to 7%. Coal-fired generation, both conventional and advanced, may increase from 51% to 57%. Nuclear generation may also increase, but estimates of its share are varied. By 2000, nuclear power will likely account for 22% to 32% of total generation, depending on how much capacity is built and licensed by the end of the century. The amount of hydroelectric power capacity is expected to increase, but hydro's share of total generation will not be as large as it is today because almost all sites have already been developed and other sources will generate greater amounts of electricity. Geothermal generation may increase from its negligible 0.2% to 3%. And about 1% of the nation's electricity will be generated by emerging solar, wind, and biomass technologies.

How are all these fuels used to generate electricity? We will find out in the next chapter.

Chapter 4

PRINCIPLES OF ELECTRIC POWER GENERATION

According to the laws of physics, energy cannot be created or destroyed; it can only be converted from one form to another. We have depended so long on a particular form of energy—the chemical energy in fossil fuels, converted to useful form by burning—that we have not prepared ourselves to undertake the large-scale conversion of other energy sources that have always been available. The problem is not finding energy sources but rather finding ways of turning the energy all around us to practical use.

THE NEED FOR CONVERSION

The kinetic energy of the wind, for example, is well suited to the work of propelling sailboats. But consider the task of toasting a slice of bread. Holding the bread in the wind will not do the work of toasting it, even if the kinetic energy of the wind were increased to hurricane proportions. In this case, the form of energy we have is totally unsuited to the type of work we want done.

Wind energy *can* be used to toast bread, but only if it is converted to other energy forms first. The kinetic energy of the wind is converted by a windmill to mechanical energy, which is converted by a generator to electric energy, which is converted by the heating element of a toaster to heat energy. So the wind energy must undergo three transformations before it finally emerges as heat energy that can toast the bread in an electric toaster.

The conversions needed to yield useful energy for any particular application vary with the form of energy we start with and the form we want to finally produce. If electricity plays a part, present-day technology almost always relies on two basic conversion devices: the electric generator and the turbine.

The Electric Generator

The electric *generator* is a device for converting mechanical energy to electric energy. The process is based on the relationship between magnetism and electricity. When a wire, or any other electrically conductive material, moves across a magnetic field, an electric current is set up in the wire. Only the relative motion of the electric conductor and the magnetic field is important. The conductor can move with respect to a stationary field, or the field can move with respect to a stationary conductor.

The large generators used by the electric utility industry have a stationary conductor (Figure 4-1). A magnet attached to the end of a rotating shaft (turned by mechanical energy) is positioned inside a stationary ring that is wrapped with a long, continuous piece of wire. When the magnet rotates, it induces a small electric current in each section of wire as it passes by.

From the magnet's point of view, each section of wire constitutes a small, separate electric conductor. But the individual sections actually make up one long wire, and all the small currents add up to one current of considerable size. This current is what the utility uses to fill the public's demand for electric power.

The Turbine

The energy conversion device that provides the mechanical energy needed to turn the shaft that rotates the generator's magnet is called a *turbine*. It converts the kinetic energy of a moving fluid (either liquid or gas) to mechanical energy.

WIND TURBINES   The simplest type of turbine is the common windmill, or wind turbine. When wind blows

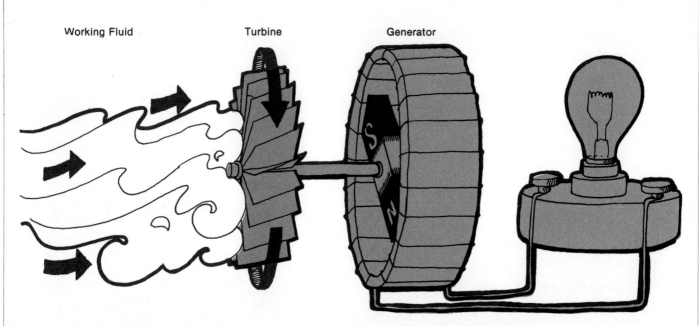

Figure 4-1. Turbine Generator Combination
This simplified drawing shows the relationship between the three essential ingredients in conventional electric power generation: the working fluid, the turbine, and the generator. What occurs is the conversion of the fluid's kinetic energy to mechanical energy in the turbine and finally to electric energy in the generator.

against the blades of a wind turbine, it exerts a force on them that causes them to rotate like a pinwheel. The rotating blades pull the shaft to which they are attached around with them, constituting mechanical energy.

WATER AND GAS TURBINES   Other types of turbines are basically the same as the wind turbine, but they use other moving fluids—or *working fluids,* as they are called—to provide the kinetic energy. Hydraulic turbines, for example, use water as it falls from a dam to push the turbine blades. And when clean fuels such as natural gas are burned, the hot gases produced in the burning process can be used directly to turn the blades of a gas turbine.

STEAM TURBINES   With dirtier fuels—coal, for example—the heat produced by burning the fuel must be transferred to a working fluid that is cleaner than the hot combustion gases in order to cut down on corrosion of the turbine blades. The fluid of choice is usually steam. The coal or other fuel is burned to heat water in a boiler, and the resulting steam then turns the blades of the turbine. The steam turbine is by far the most common type of turbine used in electricity generation today.

CONVERSION EFFICIENCY
Energy can be neither created nor destroyed, only transformed. One would expect, then, that when we convert the chemical energy in coal to electric energy, we would end up with as much energy in electric form as we started with in chemical form. Unfortunately, this is not the case. During any process, energy moves to a more random, less concentrated state. This means that whenever energy is converted from one form to another, we always end up with a smaller amount of *useful* energy.

The Sawdust Principle
Suppose you have a wooden board that you want to convert to another form—say, two smaller boards. If you saw it in half, you end up with two boards that together add up to *almost* the same amount as you started with. But a small amount of board has been changed to sawdust in the conversion process, and this sawdust, though still wood, is not suitable in its present form for building things. Every time the board is cut—that is, during every conversion process—a little more of the board is lost.

It would appear, then, that the secret of retaining high efficiency (the ratio of what you get out to what you put in) in energy production is to make as few conversions as possible. There is some truth to this, but in practice it makes a poor rule of thumb, because the efficiencies of conversions we can actually perform vary with certain practical physical limitations built into the processes. Some processes simply produce more "sawdust" than others.

The typical generator's efficiency in converting mechanical energy to electric energy is a very high 97–98%, but energy losses during other parts of the total power generation process reduce the efficiency of the typical coal-fired power plant to 30–38%. Some other conversion processes, such as ocean-thermal conversion to electricity, fall as low as 6–7% on the efficiency scale (see Chapter 6). The values for most conversion systems, of course, lie between these extremes. The trick is to use high-efficiency conversions when possible and, when low-efficiency processes must be used, to keep efficiency as close as possible to the theoretical maximum by use of technological improvements.

How Efficient Is Nature?

The conversions occurring in nature can be quite inefficient. Consider the conversion chain in which solar energy is converted through plant growth into food, which people then eat and convert into physical work. Although this is probably the most important energy conversion of all for human beings, its efficiency is only about 0.25%. If cattle eat the vegetation and then are used for food, efficiency drops even further, to 0.025%.

CONTROLLING ENERGY LOSS

In electric power generation, the "sawdust" lost from the system is mostly heat energy. Minimizing heat loss is a major engineering challenge.

Heat Recycle

Some waste heat can be recycled within the system, carried either by hot air or by steam. In a steam turbine system, water is heated in a big boiler with a fire underneath. Much of the fire's heat is put to use in boiling the water, but a lot escapes up the flue just as heat escapes up a chimney. In a power plant, some of this waste heat is transferred to the incoming air, which is then piped back to the fuel combustion chamber to give the burning process a thermal head start.

A similar heat saving is possible by recycling the turbine's working fluid, the steam. After the steam has traveled through the turbine, it has lost a lot of heat in turning the turbine blades, but it is still quite hot. Some of this steam is then used to reheat water returning to the boiler, so that less fuel will be needed to heat it up to working levels again.

Superconducting Generation

A second approach to minimizing energy loss relies on extreme cold. All metals, including the wire coils used in an electric generator, resist the flow of electricity to some extent, and this resistance means lower efficiency. When certain metals and alloys are cooled to extremely low temperatures, however, this resistance all but vanishes.

The property that allows these metals and alloys, when cold enough, to exhibit an essentially loss-free flow of electric current is called *superconductivity*. In the new superconducting generator now under development, the wire coils will be made of a special alloy and cooled by the circulation of liquid helium to −453°F. It is estimated that this new technique will cut the generator's electricity losses almost in half, boosting its already very high conversion efficiency to almost 99%.

Full Resource Use

A third way to avoid energy loss and so improve the efficiency of power generation is to use as much as possible of the energy contained in the fuel.

For a start, the fuel must be burned properly. Coal combustion is a matter of reacting the energy-rich coal with oxygen. Pulverizing the coal increases surface contact between the coal and the oxygen, and controlling the amount of oxygen available to the coal promotes thorough burning. By use of such physical and chemical controls, a great percentage of the energy in the coal is converted to heat. Very little unburned coal containing unused energy remains.

With some conversion processes, such as nuclear fission, the processes themselves prevent full use of fuel the first time through. In cases like these, the recyclable fuel is separated from the waste and used again (see Chapter 5).

Cogeneration

A fourth approach, distribution of power generators' heat to outside systems, can also promote efficient energy use in a larger sense. The idea is to make use of the excess heat from electricity generation that would otherwise be wasted. By harnessing the extra heat energy, we can get more work out of the same amount of fuel.

The concept of *cogeneration* involves the joint production of electricity and useful heat from a common source. The heat extracted from an electricity-producing steam turbine will be of higher or lower temperature, depending on the point in the cycle at which it is removed. There are cogeneration applications for both high and low temperatures.

HIGH TEMPERATURES   High-temperature steam from the turbine can be used for district heating, where heat energy is piped to residential and commercial customers to provide space heating and hot water. The steam can also be used directly in industrial processes.

But high-temperature steam from the turbine is not really waste heat. Extracting it from the turbine for use in heating means that the generator to which the turbine is connected will produce less electricity. An increase in efficiency shows up only when we consider all the systems involved, including those outside the power generation cycle.

LOW TEMPERATURES   Low-temperature heat, available in the form of hot or warm water rather than steam, really is waste heat in the sense that it cannot be used to generate more electricity. Its present uses are mainly for aquaculture (providing warm water environments for raising fish) and agriculture (warming soil to promote faster plant growth and heating air in greenhouses).

When does cogeneration make economic sense? The physical location of the heat-using system in relation to the heat-producing system is critical, for example, because of the high cost of heat distribution networks and the problem of high heat losses over distance. Some potential applications can save energy, but others could actually waste it. The only time complete fuel saving occurs is when all the electricity and all the steam made are used, but that is not always possible.

Nonutility cogenerators, such as pulp and paper mills, chemical plants, shopping centers, and hospital complexes, usually find small oil and gas generators best suited to their scale of operation. Some air quality officials worry that this scattering of fossil-fuel electricity generators throughout urban areas could spread air pollution. Oil- and gas-powered cogeneration may also represent a detour from the path toward greater use of domestic coal to supply the nation's electric power needs.

UTILITY POWER GENERATION

The rhythm of a utility's power output depends heavily on the timing of electricity use. Tailoring the supply of electricity to the ups and downs of demand is the only way to operate efficiently. When all the consuming sectors—industry, business, homes, and transportation—are using electricity at once, the utility must be ready to meet that demand. It must also be ready to cut back when demand dips. Flexibility is essential in providing a reliable supply of power to customers at the lowest possible cost.

Load

The total *load* that the utility must supply is the sum of all its customers' demands. Because of the varied schedules on which customers use electricity, the load varies over the day, the week, and the year. Naturally, the load tends to be lowest at night, when most people are asleep, and highest during the day, when the most appliances are in use.

In warmer parts of the country, where the use of air conditioning is widespread, the highest, or *peak*, load occurs on hot summer afternoons. In cooler parts of the country, where air conditioning is less common, peak loads tend to be due to combined heating, lighting, and appliance loads in the winter.

Meeting the Load

To meet the varying system load, utilities typically use several kinds of generators. Some large utilities will have as many as a hundred units in service during the peak period. The decision as to which generators to run at any given time comes from the dispatcher's office, where all the utility's power supply operations are coordinated.

The most important factors that the dispatcher weighs in assigning load to each generator are:
- The cost of electricity from the generator
- The generator's maximum capacity to produce electricity
- Reliability considerations for the entire system
- Each generator's maintenance requirements
- Environmental considerations

COST   The cost of electricity delivered to the consumer from each generator includes both fixed and variable components. The fixed costs are those that exist whether or not the generator is actually running—for example, the cost of the generator itself. The variable costs, such as fuel expense, are those that accrue only when the generator is in use.

Some types of generators have high fixed costs in proportion to their variable costs, whereas others are relatively cheap to build but expensive to run. These characteristics help determine the order in which the dispatcher will call different units into service. This procedure is termed *economic dispatch,* and the aim is to keep the overall cost of electricity delivered to the customer as low as possible.

If a plant is expensive to build but will run on fairly cheap fuel, it makes sense to run that unit as much as possible. That way its high construction costs can be recovered by selling the electricity it produces, and customers can have a steady supply of electricity without relying on scarce or costly fuels. Such *baseload* plants, as they are called, are the ones we depend on to supply electricity around the clock. Typically, utilities use coal-fired or nuclear plants to provide baseload service, because they are far cheaper to run for prolonged periods than are generators fueled by expensive oil or gas (Table 4-1).

At the other end of the generating spectrum, oil and gas units still play a necessary role for *peaking* service—that is, to provide electricity for those few hours of the day or days of the year when demand hits its peak. This role calls for units that can start and stop quickly and frequently, a requirement that the agile oil and gas generators can fill very well in comparison with the more cumbersome coal and nuclear units. The high cost of fuel oil and gas units does not matter so much because they are used for only a few hours at a time. And their lower fixed costs mean that oil and gas generators do not have to run as much in order to pay for themselves over time.

Between the extremes of baseload and peaking demand falls the *intermediate* load or the portion of the utility load that varies daily. Intermediate plants start up every morning and are turned down every night at about the time that most people go to bed. Running more hours than peaking units but fewer than baseload ones, they usually display a more balanced profile of fixed and variable costs. Utilities typically meet intermediate load with older units that once functioned around the clock but have now been replaced by newer, more efficient units for baseload service.

Throughout the day, the dispatcher "stacks" generators into a load curve as illustrated in Figure 4-2. Unit 1, the most economical, is operated at all times. Unit 2, which is more expensive to operate than Unit 1 but less expensive to operate than all others, is used except during the period 1 A.M. to 6 A.M., when it is backed down because the system load is at its lowest

Table 4-1. Typical Generating System
This table shows why, in cost terms, the six generators are dispatched in the order that they are. Unit 1 is the cheapest to operate, and Unit 6 is the most expensive. Notice that the baseload Units 1 and 2 are bigger than the others and operate on coal or nuclear fuel, both of which cost less per kilowatthour of power generated than the oil used to fuel Unit 6.

Unit	Capacity (megawatts)	Fuel	Operating Cost (cents per kilowatthour)
1	1000	Nuclear	0.9
2	800	Coal	1.6
3	300	Coal	1.8
4	150	Oil	5.4
5	100	Oil	5.7
6	50	Oil	6.3

level. The other generators, Units 3, 4, 5, and 6, are added in order of operating cost, so that Unit 6 (the most expensive to run) is used as little as possible.

CAPACITY   A generator's maximum capacity is the maximum power it can provide. This capacity is normally measured in *megawatts* (millions of watts, or thousands of kilowatts). Most generators can be operated at any level of output between about 20% and 100% of full load, although operation at the low end of this range would be quite inefficient.

RELIABILITY   Reliability is also of great concern to the dispatcher, who must provide for the possibility that one or more generators may fail to produce power at any time. This is done by ensuring that other generators are available on short notice. On large systems the dispatcher will operate several units partially loaded at all times, so they can be brought to full load quickly if they are needed. Most generators require a warmup period and cannot produce instant power the moment they are turned on.

An excess of capacity beyond the actual load is termed *reserve,* and the *reserve margin* is the percentage of generating capacity available in excess of the peak load. Most utilities have found that in order to maintain what they view as adequate reliability, they must have a 15–20% reserve margin; that is, they must maintain 15–20% more capacity than is usually needed so they can still provide reliable service in case some units break down or demand suddenly surges.

MAINTENANCE   Besides occasional repairs, generators need regular maintenance. Nuclear units must be refueled every 12 to 18 months, for example, and fossil fuel units need boiler maintenance for several weeks each year. The dispatcher must allow, then, for the scheduled absence of each generator. Most maintenance is scheduled during seasons when the load is relatively low—typically in the spring and fall.

ENVIRONMENT   Increasingly stringent environmental regulations have forced some utilities to deviate from the economic dispatch procedure described above. In southern California, for example, on very smoggy days generators are dispatched in an order designed to minimize air pollution even if it results in higher dollar costs.

Costs of Providing Electricity

EQUIPMENT   The major component of the cost of electricity is the investment that the utility makes in equipment for generation, transmission, and distribution. The total value of this equipment is termed the utility's *rate base.* Just as bank depositors expect to earn interest on the amount they have deposited in their accounts, so the investors who provide money for the utility's equipment require a return on their investment, and they are paid a fixed proportion of this rate base every year.

FUEL   Another big part of the cost of electricity is fuel, an expense that the dispatcher attempts to minimize by economic dispatch. In the load curve illustrated in Figure 4-2, the expenditure for fuel for each unit is the area shown for that unit (or energy supplied by that unit), in kilowatthours, multiplied by that unit's fuel cost per kilowatthour. As is apparent, the generating units with higher fuel costs are in operation during peak periods of the day. This fact has encouraged many utilities to apply for time-of-use rates, which would reflect the different costs of providing power at different times of day and different seasons. Under this rate structure, customers would pay higher prices during peak periods than during off-peak periods.

OPERATION AND MAINTENANCE   Other costs the utility must bear are mainly labor. Those related to the physical system are maintenance of equipment (such as boiler maintenance once a year and turbine maintenance every three or four years) and operation of the generating units, substations, and system dispatch offices. Moreover, there are the costs of engineering, accounting, customer services, and reading customers' meters.

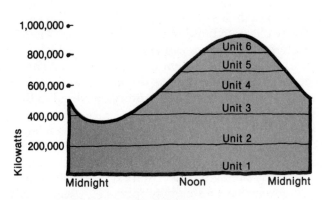

Figure 4-2. Generator Dispatch Procedure
Generators are stacked into a typical daily load curve, one with demand peaking between 5 and 7 P.M. As demand rises during the day, more and more generating units swing into action. Peaking Unit 6 operates for only a few hours before evening demand slacks, and it is backed down to await the next peak.

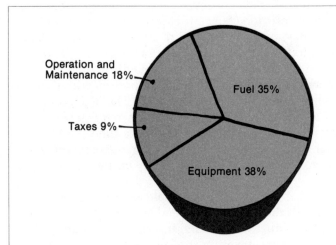

Figure 4-3. Typical Costs of Providing Electricity
Equipment still accounts for the lion's share of power generation costs, although fuel has grown increasingly expensive in recent years. Worker salaries account for most of the operating and maintenance costs, with taxes taking a substantial bite as well.

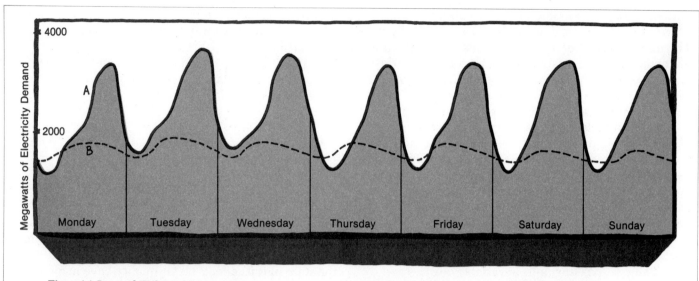

Figure 4-4. Low and High Load Factors
These two load curves reflect a typical week in August for two customer classes served by a summer-peaking utility. The residential curve (A) has a low load factor, fluctuating throughout the week and rising sharply in the late afternoon each day. The industrial curve (B) shows a high load factor, since factories operating around the clock consume electricity at a comparatively steady pace. This steadiness allows more efficient use of generating equipment than the wasteful stop-and-go pattern shown in curve A.

TAXES   Taxes add significantly to the operating costs of investor-owned utilities. These include both income taxes and property taxes. Because utilities typically have a large investment in equipment, property taxes paid to local governments average about 8% of the total cost of electricity. Government-owned utilities often make "payments in lieu of taxes" to local government bodies. A typical distribution of all these costs is shown in Figure 4-3.

Managing the Load
Because equipment is often the biggest single expense in providing electric power, efficient use of that equipment is paramount in holding down costs. This is the goal of *load management*: to promote maximum efficiency in the use of installed generating capacity. The strategy is to distribute customer demand more evenly over time, thus moderating peak demand and allowing the utility's more economic baseload/intermediate capacity to carry a greater share of the load.

LOAD FACTORS   Steadiness of use is the key. A utility's *load factor* is its average load as a percentage of its peak load. A high load factor signifies that its customers are drawing power fairly steadily around the clock, day after day. A low load factor describes the opposite condition, where customer demand is characterized by sharp peaks and valleys.

Figure 4-4 shows how the two load patterns differ in appearance. Notice that serving customer class A, with its low load factor, requires almost twice as much generating capacity as serving customer class B—but that A does *not* buy twice as much electricity. This higher fixed cost in relation to sales means that A is costlier to serve. Serving A requires additional capacity that sits idle much of the time but must still be built and held in readiness to meet A's demand peaks. Because it is so expensive to build and maintain this idle capacity, the task of load management is to smooth out load patterns like A's as much as possible.

METHODS FOR LOAD MANAGEMENT   One method already mentioned is time-of-use pricing of electric power. The purpose is to encourage customers, especially households, to shift electricity use to off-peak periods, thereby spreading their consumption more evenly over the hours, days, and even seasons of the year. This allows the utility to meet customer demand without maintaining so much capacity strictly for peaks. Additional benefits include conservation of the oil frequently used to fuel peaking generators and protection of air quality from the pollutant buildup that can occur from burning fossil fuels during periods of peak demand.

Interruptible power agreements are another tool for leveling utility loads. In exchange for lower rates, certain customers—usually large industrial concerns—agree that their electricity can be turned down or even turned off when power is needed elsewhere in the system to meet peak demands. Again, this lowers system generating costs by reducing the amount of capacity needed (a fixed cost) and also by reducing hours of use for expensive-to-run oil and gas peaking units (a variable cost).

Power cutoffs may also be keyed to specific appliances. In Detroit, for example, 200,000 customer water heaters are under utility radio control, so that the utility can shave system peaks by turning off some of these water heaters when the system is already going full tilt to meet a demand surge. Inconvenience is minimal, since most water heaters have enough stored hot water to meet customer needs until the power is turned on again. The point is to help avoid brownouts or blackouts during periods of very high demand.

None of these techniques is as yet widespread, and local needs must determine whether and where they are appropriate. The National Energy Act of 1978 has backed the further development of load management methods, and their use is expected to grow. The aim is to allow utilities and customers alike to benefit from the cost savings that come from smoothing out the peaks and valleys of customer demand.

A different approach to wringing greater efficiency out of installed generating equipment is the concept of energy storage by utilities themselves. This strategy, to be discussed at greater length in Chapter 7, can likewise reduce the need for and use of peaking capacity by permitting maximum use of baseload plants.

Choosing the Appropriate Technology

With the anticipation that loads will continue to grow in the future, utilities are now planning for the installation of new generating equipment. Like daily dispatching decisions, the choice of generating units to be added to the system involves a trade-off between equipment costs and fuel costs. The result in both cases

is typically a mix of technologies, since a mix is usually the most economical way to meet customer demand in a particular service area.

Besides the trade-off between fixed and variable generation costs, planners also must deal with several other factors in their attempt to minimize the cost of electricity:

• Fuel availability
• Cooling water availability
• Environmental constraints
• Lead time

FUEL   In some parts of the country, certain types of fuel have been effectively unavailable. For example, coal has not been burned for electric power in California because the cost of transporting it from mines outside the state plus its effects on air quality have made it seem relatively unattractive. California utilities have relied instead on hydroelectric sources, oil, gas, and nuclear fuel for generating electricity. But now, because the costs of oil and gas have risen and safety requirements for nuclear energy have become more stringent, coal may soon be used in California despite transportation and air cleanup costs.

COOLING WATER   Another problem that constrains utility planners in selecting sites for generators is the availability of cooling water. Ample supplies of water are necessary to condense the steam used in most electric generating units. This is a special concern in the western United States, where water supplies are sometimes barely sufficient for agricultural and other needs.

ENVIRONMENT   Other environmental constraints, imposed by a society increasingly concerned about pollution, set new limits on the way electricity can be generated. For example, limits on the emissions that power plants can release to the atmosphere—and the consequent need to invest in emissions control equipment—have substantially increased the direct cost of generating electricity from coal. Fortunately, there are new processes under development that promise better performance (see Chapters 6 and 8).

LEAD TIME   Long lead times—the years that elapse between the time that utility planners decide to build a large power plant and the time that the new plant will begin to produce electricity—are required for adding new capacity. These lead times vary according to plant type. Coal-fired plants, for example, can be added more rapidly than nuclear plants. But the average lead time is still about 10 years and is growing longer. So planning ahead is essential.

In this chapter, we have considered the scientific principles of energy conversion that allow human beings to generate electric power. We have also looked at some of the general operating principles that utilities follow in generating that power. In contrast, the next two chapters deal very specifically with actual generating technologies. Exactly how do utilities make the electric power that we all consume? What technologies provide most of today's electricity, and what alternatives may there be for the future?

Chapter 5

TODAY'S GENERATING OPTIONS

The fossil fuels—coal, oil, and natural gas—are the source of over three-quarters of the electricity generated in the United States today. The rest comes from nuclear plants and hydroelectric plants, with a fraction added by the nation's only dry steam geothermal installation. In 1980 we received about 50% of our kilowatthours from coal, 11% from oil, 15% from gas, 11% from nuclear plants, and 12% from hydroelectric plants.

POWER FROM FOSSIL FUELS

The fossil fuels come from decayed plant and animal life that was buried under layers of earth hundreds of millions of years ago. These fuels contain chemical energy that can be converted to heat energy by burning.

Steam Turbines

Most of the electricity produced in the United States is generated in steam turbine power plants. In a fossil fuel steam plant, the fuel is burned in a large vessel called a *boiler,* which is enclosed by an assembly of metal tubing (Figure 5-1). The fuel ignites as it enters the boiler, and the heat of combustion transfers to water that is circulating through the tubes. This water leaves the boiler as superheated steam and enters a steam turbine at about 1000°F. There it pushes against turbine blades that turn the shaft of a generator to create electricity. After it passes through the turbine, the steam is condensed and recirculated to the boiler in a closed loop for reheating.

This process is basically the same for all fossil fuels, although coal combustion generally requires a larger boiler than oil or gas combustion to achieve the same power output.

COAL   In coal-fired plants, the coal is crushed to powder for easy burning and then carried into the boiler in a stream of hot air. Coal can be cleaned before it is burned to eliminate gross impurities, but the standard practice is to burn the pulverized coal basically as is and to clean the resulting combustion gases before releasing them to the atmosphere. Because there are several different types of coal, varying in chemical structure and energy yield, coal boilers must be custom-designed according to coal type.

OIL AND GAS   Oil and natural gas can also serve as boiler fuels in steam power plants. Oil generally enters the boiler as a fine spray of oil droplets in hot air. Natural gas burns much more cleanly than coal or oil.

Oil and natural gas are both quite expensive compared with coal. Even with the added cost of post-combustion emissions cleanup, direct firing of coal in a steam power plant remains the cheapest form of fossil fuel power production on a large scale.

Gas Turbines

Oil and natural gas can also run gas turbines. In a gas turbine, the hot gases produced when the fuel burns in a high-pressure combustion chamber are passed directly through the turbine, which then spins a generator to produce electricity. The turbine also spins a compressor to provide high-pressure air to the combustion chamber. This cycle is fundamentally different from that of the

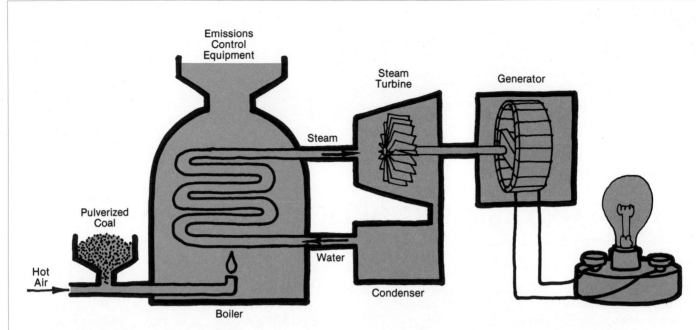

Figure 5-1. Coal-Fired Steam Power System
Crushed coal ignites as it blows into the boiler on a stream of hot air. The heat from the burning coal changes the water in the tubes into steam, and the steam turns the turbine blades that activate the generator. The system then condenses the steam to water and recirculates it to the boiler, where it is heated and used again.

steam turbine, where the heat of combustion is transferred to steam, which then serves as the turbine's working fluid.

RAPID RESPONSE   Probably the gas turbine's biggest advantage is that it can work up to speed about three times faster than a steam turbine. This rapid response makes it much more effective in providing extra capacity during periods of peak load demand (see Chapter 4). It is also much smaller and much less expensive than its steam counterpart and is thus easier to transport and install on short notice.

LIMITED CAPACITY   On the other hand, since the units are smaller, gas turbines have a much smaller power-producing capacity: a maximum of about 100 megawatts, compared with 1300 megawatts for a large steam turbine plant. This means that more gas turbines are needed for a given power output, which is definitely a disadvantage for providing baseload power on a large scale. Moreover, gas turbines run on expensive liquid and gaseous fuels.

CORROSION   Corrosion presents another snag. Whereas the only fluid that passes through the steam turbine is clean steam, the gas turbine must endure extremely hot, corrosive combustion gases. As a result, parts can wear out easily, and expensive grades of highly refined oil must be used to minimize breakdowns.

Because of its speed in responding to demand surges, the gas turbine—relatively cheap to build and install but costly to run—generally functions as a supplement to baseload steam units for satisfying peak demand.

Combined-Cycle Systems

The gas turbine, typical of peaking units, has lower efficiency than the steam turbines used for baseload power. But this problem eases when the gas turbine is coupled with a steam turbine in a combined cycle (Figure 5-2).

At about 1900°F, the combustion gases that enter a gas turbine are much hotter than the steam used in a steam turbine (1000°F). Even after they have passed through the turbine, these combustion gases are still about 1000°F. The combined-cycle system pipes the hot gases (which have already spun one turbine generator) into a "waste heat recovery" steam boiler, where they

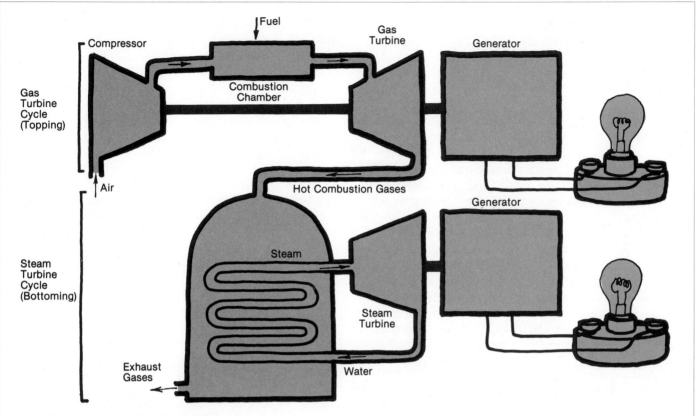

Figure 5-2. Combined-Cycle Power System
A combined-cycle system contains two separate electric generators. Fuel burned in the combustion chamber produces hot gases that first spin the gas turbine in the topping cycle. Then these same gases, which are still hot enough to boil water, make steam to turn the steam turbine linked to a second generator in the bottoming cycle.

heat water in the boiler tubes to produce steam that can produce electricity by running a second generator. So two generators are able to produce electricity from one initial fuel input.

TOPPING AND BOTTOMING CYCLES   The higher-temperature cycle (in this case, the gas turbine side) is called the *topping cycle,* and the lower-temperature cycle (the steam turbine side) is called the *bottoming cycle.* The heat delivered to the steam boiler from the gas turbine is not at as high a temperature as that normally created in a boiler by direct firing of fuel, but it would otherwise be wasted. This use of waste heat means that the efficiency of the combined cycle as a whole is higher than that of either type of turbine used separately.

FLEXIBLE COMBINATIONS   Because steam turbines are larger than gas turbines, a combined-cycle system often uses many gas turbines to feed hot exhaust to one steam boiler. The gas and steam cycles both run during normal operation, but the system is designed so that the gas turbine cycle can operate alone, making normal maintenance and repairs on the steam cycle possible without complete loss of generating capacity.

The main drawback to combined-cycle systems today is that they depend on expensive liquid and gaseous fuels. One hope for the future is to integrate combined-cycle systems with coal gasification and pressurized fluidized-bed combustion processes to create a highly efficient combined cycle that does not require oil or natural gas.

Diesel Generators

Another option for electricity production from oil is the diesel generator, which is actually a large internal-combustion engine connected to the shaft of an electric generator. This engine, rather than a turbine, provides the mechanical energy necessary to spin the generator for power production.

With about 5 megawatts maximum capacity, diesel generators are quite small. But like gas turbines, which are also relatively small, they offer great flexibility. Diesel generators can be transported easily and installed on very short notice. They also begin generating power the moment they start up, which is ideal for satisfying peak demand.

Although a number of units can be connected (either mechanically or electrically) to provide large blocks of power, the modest capacity of diesel units remains a drawback for large-scale generation. Like gas turbines, diesel generators generally supplement large baseload units during periods of peak demand. Diesel units also serve quite well in specialized applications—for example, when construction time or space is limited and when power is needed at a site that is far from any power line.

Prospects for Fossil Fuel Power

Although oil will continue to be an important fuel for electricity generation over the near term, rising costs and government policy on oil and natural gas resources focus the future hopes for fossil fuel use squarely on coal. With current methods of burning coal, however, it is both difficult and costly to meet increasingly stringent air quality standards. So most improvements in existing fossil fuel technologies and most of the new technologies now being developed—coal liquefaction, coal gasification, and fluidized-bed combustion—are part of a drive to find ways of generating electricity from coal in an environmentally acceptable manner (see Chapter 6).

NUCLEAR POWER

The nucleus of an atom is made up of particles called protons and neutrons, which are held together by very strong forces. The nuclei of certain atoms can be split by a process called *nuclear fission,* however, and this splitting releases energy that can be harnessed to produce electric power.

Fuel for Fission

Uranium is the fuel used today in the commercial production of electricity from nuclear fission. The uranium found in nature is a lopsided mixture of two *isotopes* (types of atom): uranium-238, which accounts for about 99.3% of natural uranium, and uranium-235, which makes up only about 0.7%. It is almost always the nucleus of the rare uranium-235 isotope that splits in a fission reaction.

THE CHAIN REACTION   The reaction begins when the nucleus of a uranium-235 atom absorbs a free-moving neutron. The extra neutron in the nucleus causes instability. As a result, the nucleus breaks apart—virtually explodes—into two main pieces called fission fragments, which are approximately equal in size, and two (sometimes three) free neutrons (Figure 5-3).

If one of the neutrons that breaks free during this process is absorbed by the nucleus of another uranium-235 atom, it will cause that nucleus to split and eject two more neutrons, either or both of which could cause yet another fission reaction, and so on. This phenomenon is known as a *chain reaction.* We can control it by controlling the number of neutrons from each fission that can go on to cause another fission reaction.

For uranium-235, the average number of neutrons released in fission is 2.07. Only one of these neutrons has to cause another fission for the chain reaction to be maintained. This is fortunate, because about half the neutrons fail to produce new fissions. Instead, they are absorbed by nonfissionable nuclei or are lost from the system.

PLUTONIUM FORMS   Many of these non-fission-causing neutrons are absorbed by the nuclei of uranium-238 atoms. Although this absorption does not cause immediate fission, it does have a big effect on the uranium-238 atom: it changes it into an atom of plutonium-239, which is fissionable. Because of this capacity for change, uranium-238 is referred to as a *fertile* isotope.

FUEL BONUS   Upon absorbing a neutron, the newly formed plutonium-239 fissions almost exactly like uranium-235, splitting into two main parts and emitting neutrons that can sustain a chain reaction. Consequently, once the plutonium forms from uranium-238, it is considered to be part of the fission fuel and does, in fact, contribute significantly to the release of nuclear energy. This formation of fissionable plutonium, which occurs to some degree in all nuclear reactors, is an important fuel bonus in conventional reactors. Moreover, it is the very

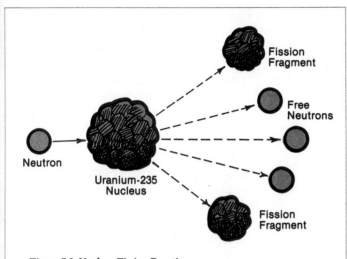

Figure 5-3. Nuclear Fission Reaction
When the nucleus of the uranium-235 atom absorbs a free-moving neutron, the nucleus breaks into two fission fragments and two or three more free neutrons. A chain reaction begins when another uranium-235 nucleus captures one of these free neutrons. Some of the kinetic energy of the fission fragments is converted into heat energy when the fragments collide with the nuclei of surrounding atoms, and it is this heat that is used to generate power.

basis for the design of the breeder reactor, an advanced reactor type that is described in Chapter 6.

HEAT GENERATES POWER   In conventional nuclear power plants, then, the main function of the neutrons released during nuclear fission is to keep the chain reaction going. The two fission fragments that are also produced in the fission process have a different function: to produce heat.

When the uranium-235 nucleus splits, the fission fragments and the neutrons carry some of the released energy with them in the form of kinetic energy. As the fission fragments collide with the nuclei of surrounding atoms, some of their kinetic energy is converted into heat energy.

It is this heat that is used to actually generate power. The heat derived from the motion of the fission fragments is transferred to water to make steam. The steam then passes through a turbine generator to generate electricity, as in a conventional coal-fired generating plant. The only real difference is in the method of heating up the water to make the steam.

Water-Cooled Reactors

The nuclear power reactors presently in commercial use in the United States are called *light water reactors* (LWRs). They have four major components: *fuel rods*, which contain pellets of uranium fuel that has been enriched to contain about 3% uranium-235 rather than the 0.7% that occurs in nature; a *moderator*, which slows down the neutrons emitted during fission so they are more easily absorbed by other nuclei; *control rods*, which contain substances that absorb neutrons readily and can thus slow down the chain reaction by grabbing the free neutrons before they are absorbed by the fissionable material; and a *coolant*, which both cools the fuel rods and carries their heat to another part of the power plant, where it is used to produce power.

Ordinary water (sometimes referred to as "light water" to differentiate it from "heavy water" used in other types of reactors) serves as both the moderator and the coolant in LWRs. The fuel rods are arrayed in assemblies that allow this water to circulate between the rods. Groups of fuel rod assemblies make up the reactor core, and the entire core is covered by the water contained in the reactor vessel. The control rods, which are dispersed among the fuel rods, can be moved in or out of the core to decrease or increase nuclear activity.

There are two major types of LWR: the pressurized water reactor (PWR) and the boiling water reactor (BWR). They differ basically in how they use the heat carried by the cooling water to drive the steam turbine.

In the PWR, the heat generated by the fuel rods is transferred to the cooling water, which circulates through the core assembly at high pressure (Figure 5-4). The high pressure allows the water to accept a great deal of heat without boiling. The water flows from the reactor vessel in a primary loop that passes through a steam generator, where the heat in that loop moves through the walls of the tubing to the water in the steam generator. Because its pressure is lower, the water in the steam generator boils, and the resulting steam is run through a secondary loop to a turbine to produce electricity.

In the BWR, the pressure in the reactor vessel is low enough for the cooling water itself to boil (Figure 5-5). The steam is then piped directly from the top of the reactor vessel to a steam turbine, where it acts as the turbine's working fluid.

There are advantages and disadvantages to both types of reactor, but these factors trade off to make the PWR and BWR economically competitive. Of the nuclear power plants now in operation in this country, about 65% are PWRs and 35% are BWRs.

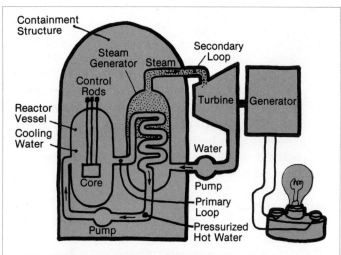

Figure 5-4. Pressurized Water Reactor (PWR)
The heat generated by nuclear fission in the reactor core transfers to the cooling water, which is then pumped in a closed primary loop through the steam generator. High pressure allows this water to carry a great deal of heat without boiling. But the water inside the steam generator vessel, kept at lower pressure, does boil, and the resulting steam in the secondary loop powers the turbine generator.

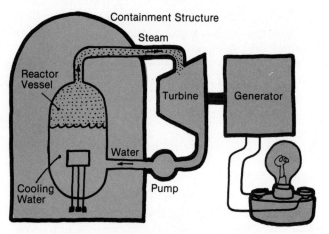

Figure 5-5. Boiling Water Reactor (BWR)
In this type of steam nuclear system, the heat from the nuclear reaction boils water in the reactor vessel itself. Steam is then piped directly to the turbine generator to produce electricity. About 35% of the nuclear power plants now operating in the United States are of this type.

Nuclear Energy Potential

Because the energy content of nuclear fuel is extremely high, just a little can go a very long way. In fact, the fission of just 1 pound of nuclear fuel releases as much energy as the combustion of about 3 million pounds of coal. This high potential for energy production makes full use of nuclear fuel particularly desirable. But nuclear changes that occur as a normal part of the fission chain reaction prevent full use.

LEFTOVER ENERGY   As more and more of the fuel's fissionable atoms split, more and more fission fragments (also known as fission *products*) are created. Like the uranium-235, uranium-238, and plutonium-239 nuclei, the fission products also absorb neutrons, but this absorption does not cause enough nuclear instability to produce fission. Soon the fission products start to present serious competition for the free neutrons, and the chain reaction slows down. One-quarter to one-third of the fissionable material in the fuel rods still remains, but it cannot be used efficiently in the ever-growing presence of the fission products, which eventually can stop the chain reaction.

REASON FOR REPROCESSING   Reprocessing the fuel can recover this unspent portion of the uranium-235 and newly formed plutonium-239 for future use. This reprocessing is one important stage in what is called the *nuclear fuel cycle*, which describes the fate of nuclear fuel materials from mining to waste disposal.

Generally, the LWR fuel cycle consists of mining and milling the natural uranium, enriching it, processing it to be used in fuel rods, fissioning the fuel in a reactor, recovering the remaining fissionable material, refabricating this material for reuse, and disposing of the remaining nuclear waste.

Although the LWRs that exist today were designed with the reprocessing and reuse of spent nuclear fuel in mind, political questions about nuclear proliferation have postponed the development of full-scale reprocessing and recycling systems. How easily could fissionable material be diverted for use by terrorists? What about military use by other nations? Until such questions are closer to resolution, establishment of commercial reprocessing plants in the United States remains unlikely.

Safety Engineering

The presence of radioactive materials in nuclear plants requires extra attention to design, construction materials, and manufacturing in addition to special control and safety systems designed specifically for nuclear power generation.

THE PHYSICS   Some safety factors are built into the physics itself. For example, the fissionable material used in the reactor is too dilute for a bomb-type reaction to be possible. Most (about 97%) of the nuclear fuel is nonfissionable uranium-238, which is no more explosive than ordinary beach sand.

AUTOMATIC SYSTEMS   Other safety features are the result of careful design. Construction and materials specifications are set so that plants are overdesigned for strength. In addition, automatic monitoring and detection systems help locate problems before they can cause serious trouble. Automatic control systems, such as the "scram" system, can then shut down the reactor in a few seconds if the monitoring devices show a potential emergency. For example, if the electricity in the plant were suddenly to go off, PWR control rods that are held over the core by electromagnets would be released, fall into the core by gravity, and stop the chain reaction.

REDUNDANCY   Such automatic systems help protect the reactor's functioning from the possibility of human error. Furthermore, these systems are *redundant;* if one safety component should fail, it is backed up by several others that take over right away—again, automatically. The emergency core-cooling system, for instance, is a complex maze of pipes, valves, and pumps whose only function is to make sure extra water can be delivered to the core fast enough to prevent overheating and melt-down in the event that regular cooling water is lost.

TRIPLE CONTAINMENT   Three separate barriers isolate radioactive materials from the atmosphere (Figure 5-6). First, stainless steel or Zircaloy tubes encase the fuel pellets. Bundled together to form the reactor core, these fuel rods rest within the steel reactor vessel. Finally, the reactor vessel and all equipment in direct contact with it are housed inside the containment building, an airtight shell of metal and reinforced concrete.

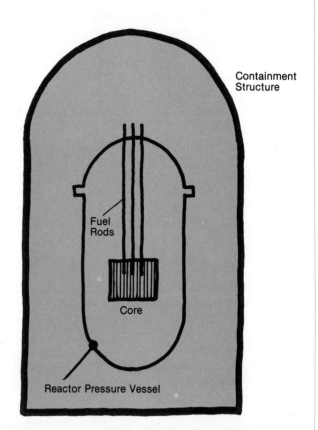

Figure 5-6. Triple Containment of Radioactive Materials (LWR)
Radioactive fuel pellets are isolated from the atmosphere by three successive barriers. Stainless steel or Zircaloy rods encase the pellets in the reactor core. These fuel rods, in turn, lie within a steel reactor pressure vessel, which is housed by an airtight containment structure made of metal and reinforced concrete.

Summing Up

On the average, nuclear power offers lower-cost electricity than any of the fossil fuels. The capital costs of building nuclear plants are high, but operation, maintenance, and fuel costs are low. Once a nuclear plant actually goes on-line, the power generated from it is relatively insensitive to inflation because fuel represents such a small portion of the final cost of nuclear-produced electricity. Nuclear plants also have a marked advantage in terms of air quality: they do not emit the atmospheric pollutants that fossil fuel plants do. This means that there is no smog problem with nuclear power generation and that expensive air cleanup is unnecessary.

The main drawbacks for nuclear power seem to be the extremely long licensing periods, which boost capital outlays by delaying construction, and public concern over nuclear safety. Various groups press for greater assurances of safety in production of nuclear energy and storage of nuclear waste. Meanwhile, the interest on the money that utilities must borrow to build the plants can soar. In view of these costly delays and concerns, many utilities have grown wary of ordering new nuclear plants, and the extent of future nuclear power growth in the United States now appears uncertain.

HYDROELECTRIC POWER

Hydroelectric power uses the kinetic energy of moving masses of water to create electricity. Energy in potential form is stored in the water behind the dam.

How It Works

Dams built across natural watercourses hold back rain and melted snow that normally flow from mountains and other high altitudes to sea level. When a dam's sluice gates are opened, water falls down passageways or conduits constructed in the dam and flows through turbines located at the bottom (Figure 5-7). As the moving water pushes against the turbine blades, the turbines spin a generator to produce electricity.

How much energy is generated depends on both the total amount of water that flows through the turbine and the height from which the water falls before reaching the turbine. This vertical distance between the water level in the reservoir and the turbine below is called *head*. High-head hydro (water held more than 500 feet above the turbine) has much more potential energy than medium- or low-head hydro.

A Renewable Resource

The water resources are completely renewable, with solar energy providing the front end of a continuing cycle. Heat from the sun evaporates water from the surfaces of oceans and lakes, and this water is stored in clouds to be returned to the earth as rain and snow. The fact that there is no fuel combustion in generating hydroelectric power has several advantages. First, power production is clean: there is no air pollution, and no significant amounts of heat are discharged into the air or water. Furthermore, almost no energy is lost as heat, and thus efficiency is much higher than in fossil fuel and nuclear plants.

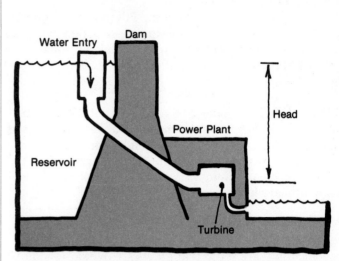

Figure 5-7. High-Head Hydroelectric System
Water running downhill from high altitudes accumulates in reservoirs created by damming natural watercourses. When this reservoir water flows through the conduit in the dam and pushes against the turbine blades, the turbine spins a generator to produce electricity.

Capital costs for dams are quite high, but operation and maintenance costs are low, and the plants have very long expected lifetimes. Dams that hold large reservoirs of water are especially valuable. They can provide both very reliable baseload power and also extra power to satisfy peak demand by simple regulation of the amount of reservoir water allowed to flow through the dam. Low-head, run-of-river plants, which have no reservoirs behind their dams, are more affected by seasonal and other variations in stream flow (Figure 5-8). Consequently, they are somewhat less reliable and have less flexibility for satisfying peak demand.

Future Prospects

Hydroelectric plants can operate only where suitable waterways are available, and many of the best of these sites have already been developed. Still, these existing plants account for only about one-third of the nation's total potential hydroelectric power capacity. More than 50% of the hydroelectric power in the United States is generated in the Pacific and Rocky Mountain states, and about 75% of the remaining undeveloped resources are also located there.

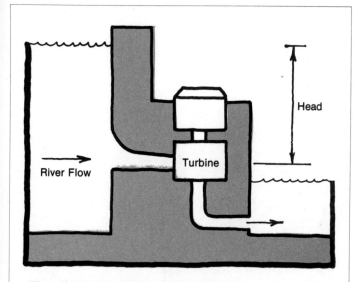

Figure 5-8. Low-Head Hydroelectric System
In a run-of-river plant, the river current rather than the force of water falling from a reservoir is what pushes against the turbine blades. Since this low-head plant has no reservoir, reliability of power output depends on seasonal and other changes in stream flow.

In the past, the heads of most undeveloped sites have been considered too low for economic power production. However, the growing scarcity and cost of fossil fuels have led to reevaluation of the economics of low-head hydro, and new turbines designed specially for low-head, run-of-river applications have been developed.

The high initial cost of the dam and associated facilities is probably the biggest concern in deciding whether or not to develop a possible site. Additional smaller dams are often required to complete the reservoir complex. And because the topography of each site is unique, standard designs can seldom be used. Each dam complex must be custom-designed.

Land costs and rights are also a factor. In fact, investment and legal considerations (sometimes concerning the conflicting rights of more than one county or state) have been so problematic that most hydroelectric projects are sponsored by the government rather than by private investors. It is likely that this pattern will con-

tinue as the nation works on new ways to squeeze the untapped power out of the lower-grade hydro resources that still remain for development.

The potential collapse or failure of dams poses a constant concern. Also, damming rivers causes ecological change, and concern about these changes has delayed construction of dams in several instances.

DRY STEAM GEOTHERMAL POWER

Geothermal power comes from heat energy buried deep beneath the earth's surface. Most of this heat is concentrated at depths beyond the reach of current drilling methods. But, in certain thermally active areas, *magma* (molten rock) from these deep regions finds its way nearer to the earth's surface through faults and fractures in the earth's crust. The heat carried by this magma transfers to rock near the surface, and the rock becomes a reservoir of recoverable heat (Figure 5-9). Most geothermal reservoirs are located in the western third of the United States.

There are four basic forms of geothermal energy: dry steam, hot water (or wet steam), geopressured resources, and hot dry rock. Of these, only dry steam represents a technology already in commercial use. (See Chapter 6 for discussion of the other three.)

When water held in the hot porous rock of underground reservoirs boils, that water becomes dry steam. This steam moves under pressure through faults to the earth's surface, sometimes bursting forth in spectacular natural geysers. There it can be harnessed and run directly through a conventional steam turbine generator combination to produce power (Figure 5-10).

Because of their inherently slow startup time, dry steam geothermal plants are best operated as baseload units. Additional generating capacity is now being added at The Geysers in California, the nation's only dry steam geothermal plant, to take full advantage of this inexpensive fuel that the earth provides. But significant geothermal growth in this country will require development of another, more plentiful source of geothermal energy—hot water resources, which are about 20 times more plentiful than dry steam.

Today, geothermal power makes only a token contribution to our energy needs. It supplies a scant 0.2% of the nation's electricity. That role is expected to grow substantially, however, as new geothermal technologies that are still in the development stage gradually mature and join the U.S. power production lineup.

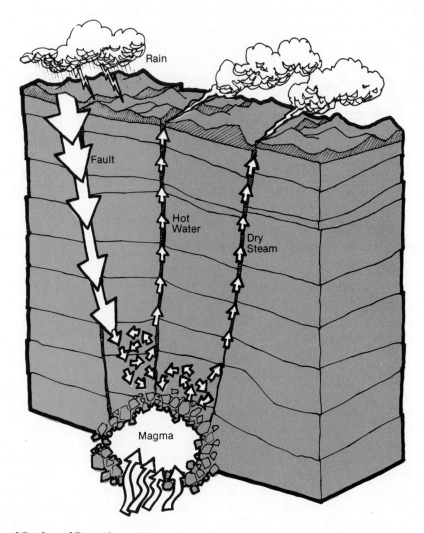

Figure 5-9. Formation of Geothermal Reservoirs
Magma—molten rock—moves from the depths of the earth up through faults in the earth's crust. When heat from this magma transfers to rock closer to the surface, the water held in the rock becomes very hot. Under intense pressure, some of this super-heated water becomes dry steam that shoots from the earth in natural geysers.

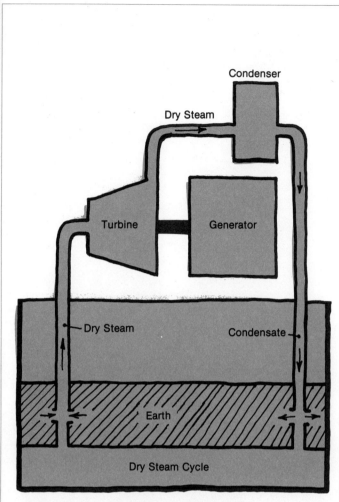

Figure 5-10. Dry Steam Geothermal Power Cycle
Dry steam from natural underground reservoirs can be piped
through a conventional steam turbine to generate electricity. The
system then returns the condensed steam, now water, to the earth.
Because the fuel is free and the equipment uncomplicated, this
method of producing electric power is relatively cheap, but dry
steam resources are unfortunately quite rare.

Chapter 6

FUTURE GENERATING OPTIONS

Technologies such as conventional fossil fuel, conventional nuclear, hydroelectric, and dry steam geothermal are already contributing to today's electric power needs. In contrast, the options discussed in this chapter are not yet commercially available. Currently, the energy and cost they require are greater than the energy and cost savings they yield. But their potential is very great. They could well play an important role in releasing the nation and the world from today's growing energy bind.

CLEAN COAL COMBUSTION

The new coal technologies endeavor to satisfy two basic needs: the need to avoid the environmental cleanup problems associated with conventional coal combustion and the need to produce coal-based fuels that can supplement or replace scarce oil and gas.

Coal Conversion

Coal conversion processes clean up coal before it is burned. Depending on the process used, coal conversion offers clean fuel in solid, liquid, or gaseous form, giving coal a new and needed versatility.

COAL LIQUEFACTION In *liquefaction*, coal changes from solid to liquid form, with impurities being removed in the process (Figure 6-1). First, the coal is crushed, dried, and pulverized. A solvent is added to the coal to form a *slurry* (a suspension of the coal particles in the liquid solvent), and the slurry is then heated and pressurized in the presence of hydrogen to dissolve the coal.

Heating the slurry triggers chemical reactions that yield various liquid and gaseous by-products. These can be removed by physical separation processes. Mineral matter and organic solids are also removed, and the solvent is recovered for reuse.

What remains is a low-sulfur liquid fuel that can be burned in existing oil-fired power plants without major plant modification. Moreover, liquefied coal can probably be refined for use in gas turbines (to provide the quick starting needed to meet demand peaks), and it may even substitute for petroleum-based diesel fuel in many applications.

Commercialization of coal liquefaction technologies is a very strong prospect. Interest in liquefaction springs from the idea that soon we may need greater quantities of liquid fuel than oil can provide at any bearable cost.

Much liquefaction research focuses on three processes: the solvent-refined coal (SRC) process, the Exxon Donor Solvent (EDS) process, and the H-Coal process. SRC-I, a clean-burning *solid* fuel, was the first product of this research. Although chemically and physically quite different from coal, this solid synthetic can substitute for pulverized coal as a boiler fuel with only minor adjustments in conventional coal plants.

Development of SRC-II, a clean-burning *liquid* fuel, is progressing as an extension of SRC-I research. The EDS and H-Coal processes differ from SRC-II primarily in that they both use chemical agents to promote and control the chemical changes that take place during *hydrogenation* (reaction with the hydrogen). With H-Coal, the coal slurry itself is hydrogenated by means of a catalyst. In the EDS process, the solvent is catalytically hydrogenated as a separate step.

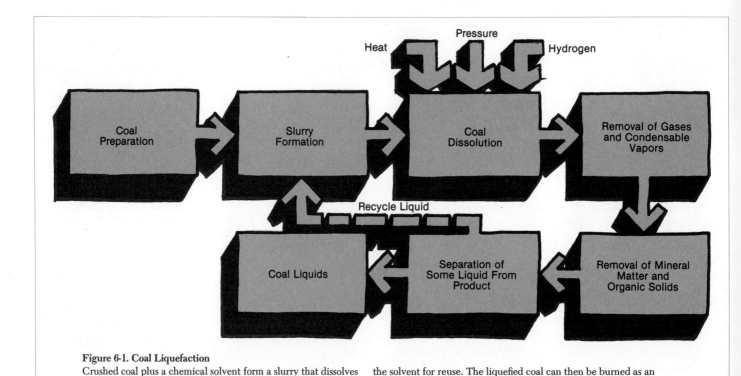

Figure 6-1. Coal Liquefaction
Crushed coal plus a chemical solvent form a slurry that dissolves under high heat and pressure in the presence of hydrogen. Separation processes remove unwanted by-products and recover the solvent for reuse. The liquefied coal can then be burned as an oil substitute in regular oil-fired power plants with only minor plant modification.

These three liquefaction routes promise similar performance and face many of the same technical barriers, although the SRC-I and SRC-II technology may advance to full-scale demonstration earlier than the others. A major barrier is environmental impact. Even though the end product is clean fuel, coal conversion itself can release significant quantities of phenols, hydrocarbon vapors, dust from coal piles, nitrogen oxides, and other pollution by-products and residues. Builders and operators of coal conversion plants must ensure the safety of plant workers as well as that of the environment. So the problems surrounding synthetic fuel manufacture from coal are compounded by questions about the nature and cost of the safety measures and environmental cleanup that must accompany such operations.

Pilot plants for all three liquefaction processes are presently in operation. If these processes reach the demonstration phase by 1985, they could together yield 450,000 barrels a day of liquid fuels by 1990 and 950,000 barrels a day by 1995.

Synthetic liquid fuels made from coal are not expected to be cheap. And because first-of-a-kind plants are usually significantly more expensive than the second or third plants built, the move to commercial production is particularly risky.

COAL GASIFICATION   Coal *gasification* is a method of producing a clean, burnable gas from coal. The basic chemistry involves oxidizing the carbon in the coal—that is, uniting it with oxygen, using either pure oxygen or the oxygen in the air (Figure 6-2).

When complete oxidation of carbon occurs (as when pulverized coal is burned conventionally), nearly all the chemical energy of the coal is released as heat energy and the remaining gases will not burn. By restricting the amount of oxygen available to the coal, however, and controlling the temperature and duration of the reaction, we can ensure only partial oxidation and form gases that will burn (mainly hydrogen and carbon monoxide).

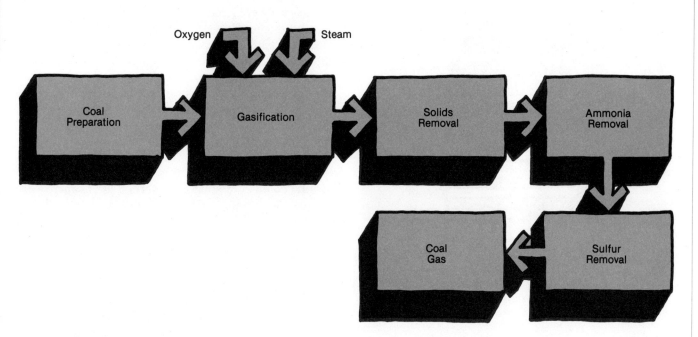

Figure 6-2. Coal Gasification
In this process, the crushed coal is partially oxidized. The result is a combustible gas. After the various by-products are removed, the clean coal gas can be burned to fuel a steam turbine or piped directly into a gas turbine.

Partial oxidation means that less energy escapes from the system in the form of heat. Much of the chemical energy originally held by the dirty coal ends up as chemical energy held by the clean, combustible gas. The gas can then fuel a boiler driving a steam turbine or work directly in a gas turbine to generate electricity.

One gasification technology, the Lurgi system, has been in commercial use for some years. There are now several second-generation, or advanced, gasifiers at the pilot plant stage, but they are not expected to become commercial until the mid-1980s. Basically, the advanced systems differ from one another in how the coal and oxidant are brought into contact and how the physical conditions (temperature, pressure, and so on) of the reaction are controlled.

Advanced coal gasification will be developed for commercial use by the mid-1980s. Cost predictions suggest that gasified coal will be able to compete with the more expensive natural liquid and gaseous fuels for providing intermediate and peak load electricity service. But this may not be its most significant application.

Tests show that when a gasification unit is integrated into a gas-steam combined cycle, the efficiency and economics of the system improve to a point competitive with conventional coal-fired plants for providing base-load service. If air quality standards tighten in the future, the gasification–combined-cycle (GCC) plant will become an even more attractive option. At the current rate of research and development, GCC baseload systems could be commercial by 1990.

Fluidized-Bed Combustion

Fluidized-bed combustion (FBC) limits stack emissions by modifying and controlling the chemical reactions that occur during coal burning itself. Undesirable gases that could escape to the atmosphere either are not formed or are converted to other compounds during combustion. The FBC process allows clean burning of regular, unconverted coal.

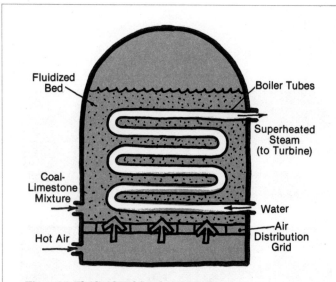

Figure 6-3. Fluidized-Bed Combustion Boiler
Coal and limestone particles mix in the turbulent hot air. As the coal burns, water boils in the boiler tubes, producing steam that runs the turbine. The limestone removes sulfur gas by-products that would otherwise go up the stack by reacting with them to form calcium sulfate, a disposable solid waste.

The crushed coal burns in what is known as a *fluidized bed:* a concentrated suspension of crushed limestone particles in a flow of hot gas, usually air (Figure 6-3). Although the bed is composed almost completely of solids, the movement of hot air causes the particles to boil like a turbulent fluid. This movement allows excellent mixing of air, coal, and limestone and promotes even distribution of heat throughout the bed.

As the coal burns, the limestone reacts chemically with the sulfur oxides normally created during coal combustion to form calcium sulfate. This waste is collected in the form of a dry, granular solid. Thus, because the sulfur oxides that can threaten air quality are captured during the actual coal burning, FBC systems do not require expensive cleanup of stack gases, although they still produce solid waste that must be disposed of properly.

The FBC process will allow us to burn low-grade and high-sulfur coals that have, in the past, been considered too dirty to use. In addition, waste fuels, such as forest residues and municipal solid waste, are more easily burned in FBC than in conventional combustion systems.

FBC systems are also easier to build than coal plants, because they are smaller and because more parts can be mass-produced. However, even though the remaining hurdles to FBC use mostly involve technical refinements, these drawbacks are by no means trivial. Ultimately—as with most new technologies—the future of FBC will depend on how its performance and economics compare with those of existing systems. Operation of the fluidized-bed combustor at high pressure makes it possible to integrate the boiler into a combined-cycle system and achieve higher efficiencies. If federal air quality standards continue to tighten, this new option for clean coal burning could come into commercial use as early as 1985.

SOLAR POWER

Solar technologies convert the light energy of the sun into other energy forms that we can control and use selectively. The natural conversion of light energy occurs all around us—in the heating of the earth's surface and atmosphere, the creation of winds and ocean waves, and the growth of plant life. It is theoretically possible to reclaim energy from these secondary sources of solar energy as well as from the sun itself, and such solar-derived energy prospects will be dealt with in later sections of this chapter.

In common use, the term *solar power* refers specifically to the conversion of sunlight into heat or electricity. Solar collectors that actively convert sunlight into heat for heating air and water in individual buildings are already available, although they are expensive. And passive solar heating occurs when a building is sited, constructed, and insulated so as to take maximum advantage of the sun's rays. The major effect of passive solar heating will be on new structures rather than on existing ones, so it could take a decade or more for the effect to show nationally. Building codes that ignore solar energy potential and a slow return on the building owners' investment have tended to hold back the more rapid adoption of passive solar heating systems.

Promising though these systems are for providing heat, neither solar collectors nor passive solar construction can meet many of the home and business energy needs normally filled by electricity. Neither, for example, can light a lamp, run a refrigerator, or power a computer. Customer-owned solar heating systems generally require electricity backup to meet these other needs.

If sunlight itself were readily convertible into electricity, solar energy could become a more versatile and hence more widely usable energy source. The two technologies being developed for this purpose are solar-thermal conversion and photovoltaic conversion.

Solar-Thermal Conversion

When a particle of light strikes a surface, the surface gives off heat. The basic idea of solar-thermal conversion is to transfer this heat to a working fluid (such as steam or air) that can then turn a turbine generator to produce electricity. Because sunlight naturally reaches earth in such a diffuse state, it is necessary to concentrate the light from a large area on the working fluid to heat it sufficiently for electricity generation. There are two schemes for achieving this concentration: the dispersed collector concept and the central receiver concept.

DISPERSED COLLECTORS   With dispersed systems, each concentrating unit works individually to convert solar energy into heat. Parabolic mirrors (mirrors in dish or trough shapes) track the sun across the sky and concentrate its light on a focal line that runs through all the mirrors (Figure 6-4). The heat concentrated by each mirror is transferred to the working fluid as it is piped through this focus, and the fluid is then pumped to a turbine generator.

CENTRAL RECEIVERS   The central receiver concept, now considered the best for economic solar-thermal conversion, uses two-axis tracking mirrors instead of parabolic ones. These *heliostats,* which are positioned around a central tower, track the sun so that the light reflected from each mirror enters a thermal receiver at the top of the tower (Figure 6-5). The heat then transfers from the thermal receiver to the working fluid, which is piped down the tower to a turbine for electricity generation. This central receiver system is also known as the *power tower* concept.

COSTS AND SITING   The main block to solar-thermal electricity is the same as for the other solar power options: it is expensive, both to construct generating systems and to ensure their reliability (the ability to provide power whenever it is needed).

The central receiver system, which is expected to be the lowest-cost solar alternative for utility use on a large scale, is still considerably more expensive than coal or nuclear power generation. The major factor here is the high cost of the heliostats and their support structures. In addition, the fact that the sun shines only during the day, and very little on some days, means that a solar

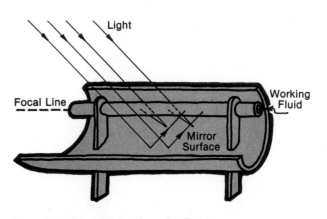

Figure 6-4. Dispersed Solar-Thermal Collector
The curved mirror tracks the sun across the sky and concentrates its rays on the focal line. The sun's rays heat the working fluid piped through this focus, and the heated fluid is then pumped to a turbine generator to produce electricity.

generating station can operate only about one-half to one-third of the time.

Most central receiver plants will probably be located in areas that receive the most direct sunlight with the least cloud cover, notably the arid Southwest. Further, a typical system needs fully a square mile of heliostats to collect enough sunlight for power generation. These siting considerations alone could restrict power towers to a relatively small portion of the country.

STORING SUNLIGHT   When the sun is shining, the fuel for generating solar power is virtually free. At night, however, this solar fuel is absolutely unavailable. Therefore, some form of storage is necessary if solar power is to be a reliable source of heat or electricity around the clock (see Chapter 7).

Low- and medium-temperature solar heat can be stored in water tanks or rock beds. Fusible salt storage technology, which is still under development, promises to increase these storage capabilities. But all such storage systems take up a great deal of space. Coal provides 10,000 to 12,000 Btu of chemical energy storage per pound, whereas water can store only about 120 Btu of thermal energy per pound.

Fortunately, our greatest need for energy occurs during the day, when the sun is shining. Converted into electricity, this sunlight can help meet utilities' intermediate power demand—the demand that picks up each

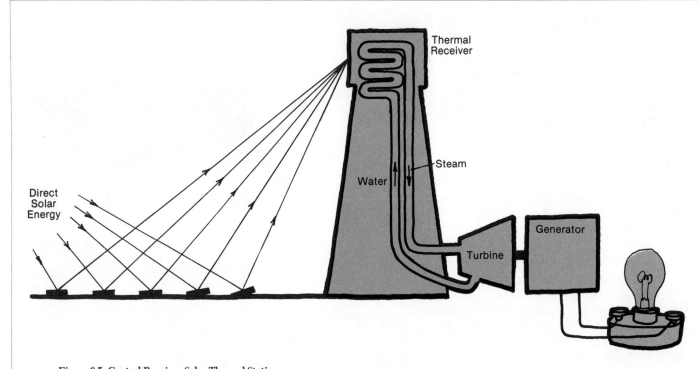

Figure 6-5. Central Receiver Solar-Thermal Station
Flat mirrors are positioned around the tower so that the rays reflected from each mirror focus on the thermal receiver. The heat concentrated in the receiver then transfers to the circulating water, which turns into steam that will activate the turbine generator. This power tower concept is currently the leading option for generating solar electricity.

morning and falls off each night—as well as meeting daytime demand peaks. For such intermediate and peaking service, no storage is necessary. So solar electricity could begin making a contribution to the nation's power needs even before the storage problem is solved.

Solar-thermal research presses forward, especially on the power tower concept. A 10,000-kilowatt demonstration plant that uses a steam turbine cycle is expected to be in operation by 1982. And more advanced demonstration plants that will use air, molten salt, or liquid metal as the tower's working fluid are planned for the mid-1980s.

Photovoltaics

Photovoltaic conversion generates electric power directly from sunlight rather than using the sun's heat. The electricity-generating unit is a *photovoltaic cell,* or solar cell. It is made from a *semiconductor* material, such as silicon, whose electrical conductivity can be boosted greatly by exposing it to heat, light, or voltage.

When sunlight of sufficient energy is absorbed by a semiconductor material, an electron is dislodged from one of the material's atoms (Figure 6-6). Special treatment of the semiconductor surfaces during manufacture makes the front surface of the cell more receptive to free electrons, and the electron naturally migrates in that direction. When many electrons, each carrying a negative electric charge, travel toward the front surface of the cell, the resulting imbalance of charge between the front and back surfaces constitutes a potential electric current. The front and back surfaces of the cell become charged like the negative and positive terminals of a car battery, and when they are linked, a flow of electricity occurs.

PORTABLE POWER   Several aspects of photovoltaic conversion are quite appealing. Conversion from sunlight to electricity is direct, so bulky, complicated turbine generator systems are unnecessary. Unlike solar-thermal systems, solar cells can use diffuse sunlight without concentration, and therefore could be used in places that do not receive much sunlight. Also, although large systems will probably use large arrays of cells, an individual cell delivers the same amount of electricity whether it works alone or in an array. This means that solar cells can perform as efficiently in small, dispersed applications as in large systems.

EXPENSIVE CHIPS   The major disadvantage of solar cells is their extremely high cost. The cells are so expensive to manufacture that electricity production using photovoltaics costs between 10 and 20 times as much as production by conventional power plants. The high cost of refining silicon to the degree of purity required for solid silicon cells has prompted researchers to examine the use of a cheaper substitute: thin films of semiconductor materials deposited on bases of plastic or glass. A technological breakthrough in this area could cut costs and considerably boost the prospects for photovoltaic conversion.

THE TPV CONVERTER   A solar cell variation, the *thermophotovoltaic converter* (TPV), also offers hope for solar technology, not by lowering cell costs but by dramatically improving cell efficiencies. With this concept, a metallic radiating element in front of the solar cell is heated by concentrated sunlight and as a result radiates light that is primarily in the infrared range—the light to which solar cells are most responsive. The cell thus receives a much higher percentage of the type of light it can actually use to produce electricity. The system also employs reflective coatings to recycle light that is not used on the first pass through the cell. Estimates of possible efficiencies for the TPV concept run as high as 50%, although the technology is still experimental and its future highly uncertain.

SOLAR CELL PROSPECTS   Photovoltaic cells have actually been used to supply electricity in the NASA space program and in a few other special situations where power has been needed in places far from the nearest power lines. But large improvements in cost must occur before photovoltaic cells can compete with existing power generation systems on a large scale. Storage, too, remains a hurdle (see Chapter 7). We must learn how to store solar energy before it can meet reliability standards for baseload power generation, although nonbaseload applications could come much sooner.

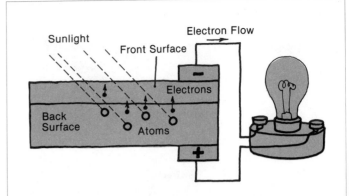

Figure 6-6. Photovoltaic Cell
The cell, made of semiconductor material such as silicon, absorbs sunlight. This action dislodges certain electrons, which migrate toward the specially treated front surface of the cell and so create an electron imbalance between the two surfaces. When the negatively charged front surface and the positively charged back surface are joined by a conductor, a flow of electricity occurs.

WIND POWER
The kinetic energy of the wind can turn a wind turbine and thus generate electricity. This kinetic energy comes mainly from solar heating of the atmosphere. The sun's rays create temperature and density differences in air masses, which, in turn, produce wind.

The Wind Turbine
A basic wind turbine consists of propeller (rotor) blades mounted at the top of a tower and connected by gears to a drive shaft that translates the rotation of the blades to a generator (Figure 6-7). The amount of energy a wind machine can convert depends on both the wind velocity and the length of the rotor blades.

An alternative to the conventional design is the vertical-axis wind turbine, which has eggbeater-type blades attached directly to a vertical shaft (Figure 6-8). The vertical-axis turbine can accept wind from any direction and therefore does not require mechanisms to turn the rotor into the wind.

Variable Wind Speeds
Although the fuel for a wind power plant is free and available in a vast range of geographical locations, it is not of uniform quality. Wind speeds below about 8 miles per hour are not sufficient to produce power with present technology; at the other extreme, powerful gusts can damage mechanical equipment and cause

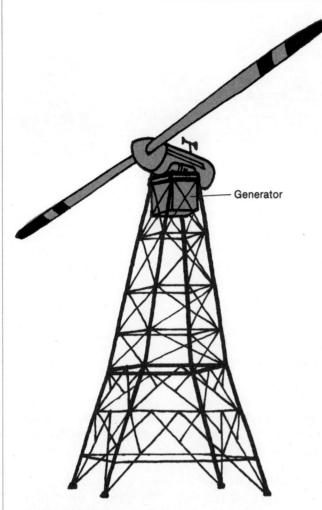

Figure 6-7. Horizontal-Axis Wind Turbine
Wind turns the propeller blades, which are connected to a drive shaft that spins the generator. The amount of electric power that a wind turbine can deliver depends on the length of its blades and the force of the wind itself.

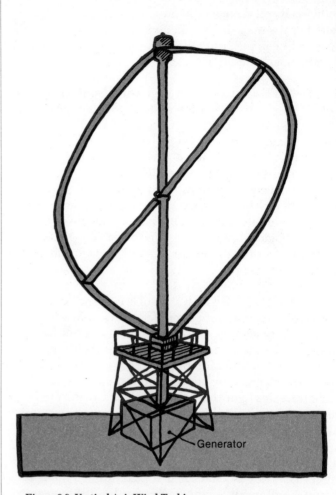

Figure 6-8. Vertical-Axis Wind Turbine
In this variation on the conventional wind turbine, eggbeater-type blades are attached directly to the shaft that drives the generator. One advantage of this turbine is that it can accept wind from any direction, but so far it is not as efficient as the conventional wind machine in generating electricity.

rotor blades to fail. Wind speed variation may cause problems in integrating wind-generated electricity with the absolutely regular current that is standard for electric systems in the United States.

Siting is very important. The Great Plains and the East and West Coasts offer the most promise for sustained, usable wind velocities. Although the operating costs for wind systems are expected to be quite low, the cost of the power plants themselves is high, considering

that it would take an 1800-square-mile area of wind machines to produce the same power produced by one large coal or nuclear plant. As other fuel costs rise, however, wind energy looks increasingly attractive in those areas where high winds prevail.

The Storage Question
Since we cannot count on a steady supply of wind energy, development of an efficient energy storage system is necessary before we can use wind for baseload electricity generation (see Chapter 7). The most cost-

effective use of wind power today is as a supplement to more conventional generating systems. Like solar power, wind power holds promise for meeting intermediate and peaking electricity loads in those parts of the nation where nature permits.

Projections of wind technology growth and large-scale commercialization are extremely uncertain. A contribution of one quad of electricity could come as early as 1995 or as late as 2015.

OCEAN POWER

The ocean offers several forms of energy that are capable of being converted to electricity: ocean-thermal energy, wave energy, and tidal energy. These energy resources are free and completely renewable, but the energy content per unit area is quite low and the cost of the power plants is very high.

Ocean-Thermal Energy Conversion

Ocean-thermal energy conversion (OTEC) generates electricity by exploiting the temperature difference between the ocean's relatively warm surface water, which is constantly being heated by the sun during the day, and the cold water more than a thousand feet below the surface.

HOW IT WORKS   One typical version of an OTEC plant consists of a cylinder about 800 feet in diameter and 1500 feet high that floats vertically, with the top just above the surface of the water. A low-boiling-point gas, such as ammonia, is pumped to the bottom of the cylinder, where it is condensed to liquid form by the cold water in a large heat exchanger. The liquid is then pumped back to the surface, where the heat of the surface water (again, in a large heat exchanger) causes it to boil. The expanding gas turns a turbine generator to produce electricity. After the gas leaves the turbine, it returns to the colder depths to repeat the cycle.

An alternative plant design pumps the cold water up to the gas for condensing rather than pumping the gas down to the cold water. Although the gas expansion-condensation cycle is technically the same, the OTEC plant can be much smaller with this variation.

LOW EFFICIENCY, HIGH COST   The larger the temperature difference between the water at the top and bottom of the cylinder, the more economical it is to produce electricity. The basic snag with OTEC is that the ocean's temperature differences are rather small—about 35°F at best. So the maximum efficiency of the whole system is only about 3% or 4%. The extremely large, costly OTEC structure can produce only relatively small amounts of power.

Good sites with a large enough temperature differential are limited. Also, since an OTEC plant would have to operate 20 to 150 miles offshore to obtain the necessary water depth, a long, costly underwater cable would be needed to transmit electricity to land. Engineers must contend with formidable problems in designing a large plant that can operate reliably for 30 or 40 years in a corrosive saltwater environment, where storm-created waves are strong, and barnacles and slime can clog the heat exchangers. In view of these constraints, the cost of generating electricity by OTEC plants will likely be too high to develop the technology on a large scale.

Wave Energy Conversion

The wind transfers some of its kinetic energy to the surface of the ocean in the form of waves. The rise and fall of the waves can be converted into hydraulic pressure by mechanical compression devices, and the pressure can drive a turbine generator to produce electricity.

With one such conversion device, the waves push against a series of plates that flap back and forth as the water moves. The mechanical motion of the plates, which are hinged to small mechanical pumps, compresses the working fluid that goes to the turbine. With another device, a floating piston inside a stationary cylinder moves up and down with the surface of the water and compresses air in the top of the cylinder. A third device consists of a line of floating rafts, with mechanical pumps located at the joints that connect them. As the rafts move up and down with respect to each other, the pumps compress the working fluid that spins the turbine generator to produce electricity.

Like energy from the sun and the wind, wave motion is intermittent and consequently unreliable for baseload power generation. Wave availability also varies with wind and current patterns, so siting the devices can be tricky. And salt water can corrode the equipment. Like OTEC plants, wave systems are expensive and present major structural problems.

Efficiency of the systems depends on the size and direction of the wave movement. Experts in Great Britain, where most of the world's wave research is being done, expect efficiencies of 25–50% for large devices, considerably better than the efficiencies possible with OTEC systems.

Tidal Energy Conversion

Tidal power comes from the regular ebb and flow of ocean tides caused by the gravitational pull exerted on the earth by celestial bodies, primarily the moon. To make use of these tides, a dam can be constructed across a coastal inlet. As the tide comes in, water runs through reversible turbines in the dam, which turn generators to produce electric power.

At high tide the turbines are reversed, and power flows again as water runs out of the inlet reservoir. The difference in elevation between water levels at low and high tides determines the amount of electricity that can be produced.

Unfortunately, tidal energy is limited to coasts, and there are relatively few sites in the world—with only two or three in the United States—where the daily tidal range is great enough to justify the large cost of building the dam. The two existing large-scale tidal plants, one in France and one in the USSR, have not reported encouraging results. Because of the high capital costs and low energy availability, tidal power is unlikely to make a significant contribution to energy needs in the United States.

BIOMASS CONVERSION

Biomass is generally defined as organic material that can provide heat. In biomass conversion, the organic material is burned directly or is chemically converted to a burnable fuel. Included in the biomass category are standing vegetation, aquatic vegetation, residues from agricultural production (stalks, husks, and so on), and animal wastes. The organic portion of municipal solid waste (garbage and trash) is also considered a biomass resource, even though the organic material—mainly paper—has gone through a useful intermediate stage.

Using biomass as fuel is not a new idea. Wood was once man's primary source of fuel. But fuels with higher energy content, such as coal and oil, eventually made wood burning uneconomical. Biomass still fills certain specific roles routinely. The paper industry burns its own sawdust and wood chip waste to supply heat for the papermaking process, and farmers sometimes burn agricultural wastes to provide heat for drying crops. It is also possible to create liquid and gaseous fuels from agricultural residues, but this source is not expected to contribute widely to synthetic fuel production simply because the waste is too expensive to collect and transport. It makes much more sense to use plant and animal waste on the farm where it is produced.

Energy Farms

One proposal for increasing biomass use is to create energy farms, where crops would be grown entirely for use as fuel. The crops, most likely sorghum and sugar cane, would be high-yield vegetation with short harvest cycles. The vegetation would be harvested and converted to synthetic fuels by *pyrolysis* (heating in the absence of oxygen), *chemical reduction* (heating in the presence of water, carbon monoxide, and chemical catalysts), or *anaerobic* (bacterial) *fermentation*. Another type of energy farm would harvest fast-growing trees such as sycamore and eucalyptus, which could be burned directly.

COMPETING WITH FOOD CROPS   These concepts have a major drawback: energy crops would be competing with food crops for land, air, and fertilizer. At current price levels, the fuel produced would be worth less than the food that could be grown on the same amount of land. And the trees grown on an energy farm would be less valuable as fuel than as lumber or paper. This competition makes many of the schemes for growing fuel on energy farms impractical from an economic standpoint, at least at today's price levels.

FARMING THE WATERS   A variation on this idea is marine energy farming of lakes and coastal areas. Crops of kelp and water hyacinth would support synthetic gas production. Although aquatic farming avoids some of the problems faced by land-based farms, the technical, economic, and environmental aspects are still being assessed.

Municipal Solid Waste

Every year, American towns and cities collect and dispose of between 110 and 150 million tons of municipal solid waste (MSW). Most MSW, which excludes sewage, becomes landfill, but actually about 75% of it is burnable. This waste offers a good chance for making use of biomass materials on a large scale. Since collection is a major problem in using most biomass and since

MSW must be collected whatever its fate, the use of MSW as fuel has an initial advantage over other biomass conversion schemes.

DIRECT BURNING   The simplest way to make use of MSW is by direct incineration of the waste in a vessel that heats water to make steam. Siting is a problem, however, since a garbage-burning plant could produce quite a variety of pollutants and offensive odors in the neighboring area. Separating the different types of waste before burning can also be difficult and costly.

CHEMICAL CONVERSION   Like plant and animal waste, MSW can be converted to synthetic liquid or gaseous fuels by pyrolysis or chemical reduction. Practical systems for these conversions are still in the development stage, and more than half of the fuel produced is used up in providing heat for the processes themselves. At this point, the resulting synthetic fuels are more expensive than their natural counterparts, and the conversions involve high capital and operating costs. Still, producing synthetics from MSW represents a definite possibility for the future, especially if oil and gas prices continue to rise.

SOLID FUEL PELLETS   The most promising technology for making use of MSW for electricity is the production of refuse-derived solid fuel. This system pulverizes the waste, separates glass and metals by mechanical or magnetic processes, and compresses the remainder into burnable pellets. Whether or not this approach is economic in any particular case depends on the relative cost of the usual means of waste disposal (usually landfill), the relative cost of transportation for the waste, and the cost of weeding out the noncombustible materials.

Although it is a renewable resource, MSW offers only a small amount of energy at any particular time—at best, 4–7% of a utility's fuel needs in a large urban area. For similar reasons of availability, all the biomass resources taken together can probably contribute no more than about 5% of the nation's total energy needs. Even this amount, however, could be very valuable in relieving our dependence on oil and gas.

EMERGING GEOTHERMAL OPTIONS

Geothermal energy from the earth's inner heat is already generating power in dry steam plants (see Chapter 5). The other geothermal possibilities—hot water, geopressured zones, and hot dry rock—are still in the development stage.

Hot Water

Hot water geothermal resources, which are buried in the earth under pressure, require more complicated power cycles than dry steam systems. Some hot water geothermal power is being generated in the United States today, and these resources are expected to yield significant new supplies over the next 15 years.

ALTERNATIVE APPROACHES   In one cycle, the hot water is *flashed* (suddenly depressurized) to turn it to steam for use as the working fluid in a steam turbine. If the flashed steam has a very low salt content, it can be used directly. But more often, highly efficient separators and steam purifiers are required to prevent minerals in the steam from damaging the turbine. An alternative to the flashed-steam cycle is the binary cycle, in which the heat of the pressurized hot water is used to vaporize a second, clean fluid that will actually turn the turbine (Figure 6-9).

SITING   Some technical concerns are directly related to the site-specific nature of geothermal plants. Because different geothermal reservoirs are likely to have different geologic and chemical characteristics, the steam and water recovered from a particular plant will have a unique temperature, pressure, and mineral composition. Thus, the design of a plant will vary significantly with respect to type of power cycle, need for cooling water, susceptibility to corrosion and mineral buildup, need for emissions control equipment, and so on. Another practical concern is the fact that known geothermal reservoirs —of which there are few—are generally situated far from population centers, which means that long transmission lines and associated losses of electricity during transit are a problem. Moreover, some geothermal reservoirs are in national or state parks, which are protected from development.

Development of hot water resources is expected to clear the way for development of geopressured and

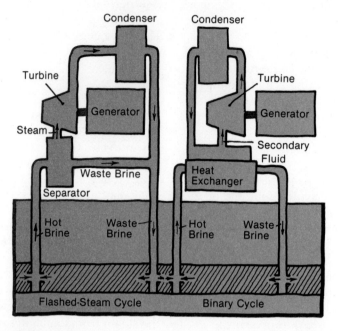

Figure 6-9. Hot Water Geothermal Power Cycles
In the flashed-steam cycle, the hot brine is flashed (suddenly depressurized) to produce steam, which is then purified and run through a conventional steam turbine to produce power. In the binary cycle, the heat of the brine is transferred by a heat exchanger to a second, clean fluid, vaporized by the heat of the brine, that actually spins the turbine generator.

hot dry rock energy systems. These more advanced forms will probably not be fully exploited, however, until the hot water technologies are demonstrated and refined.

Geopressured Zones

Petroleum exploration in the Gulf Coast region has identified reservoirs under extremely high pressure that contain hot salt water mixed with methane gas. These reservoirs, trapped in formations of impermeable shale, are referred to as *geopressured zones*. They offer energy potential in the form of both pressure and burnable methane, as well as the heat energy available from any geothermal resource.

The extent, quality, and recoverability of geopressured resources, located mainly along the coasts of Texas and Louisiana, are presently under study. The economics of recovering this resource will depend greatly on the methane content of the hot water and the scarcity of the natural gas it would supplement.

Hot Dry Rock

Extracting energy from subterranean hot dry rock means introducing a heat exchange fluid to carry the heat from the rock to the power plant. The process starts with fracturing the hot rock (using techniques common in conventional oil field and natural gas development) to provide a field of fissures through which the fluid can flow (Figure 6-10). Water is then injected deep into the fissured hot rock reservoir by a long pipe. The water rises as it heats up, and it is removed by another pipe near the top of the reservoir.

The hot water piped to the surface is used in one of the standard geothermal power cycles already described. Development of hot rock systems would increase the reserves of geothermal resources tremendously, but the technical and economic feasibility of such systems is not known, and the quantities of injection water needed may be prohibitively large.

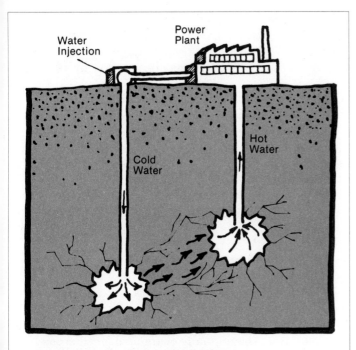

Figure 6-10. Hot Dry Rock Geothermal System
The hot rock is fractured and the cold water injected into it through a long pipe. The rock heats the water, which is then used in a standard hot water geothermal power cycle to generate electricity.

Constraints on Development

Although geothermal power is forecast to grow in the future, development of known resources is not keeping pace with new geothermal resource discovery. There are several reasons besides site restrictions for this lack of commitment.

First, geothermal heat is a relatively low-grade resource. The temperature and pressure of geothermal steam and water are low compared with the temperature and pressure of working fluids used in more conventional power plants. This means that energy conversion efficiency is also low, even with the specially designed low-pressure turbines used in geothermal plants.

Government regulation of land leases and land construction permits is also a problem, with state and federal agencies often disagreeing on regulatory policy. Questions of who owns the property and the mineral rights are often difficult to resolve.

FUEL CELLS

Fuel cells are devices that convert the chemical energy of fuels directly into electricity. This is accomplished by oxidation of the fuel inside the cell. Since no fuel is burned, far less waste heat and fewer emissions are produced than with combustion equipment. The main by-product of fuel cell energy conversion is plain water.

A Power Sandwich

The cell consists of an *electrolyte* (an electrically conductive chemical medium) sandwiched between two electrodes (Figure 6-11). Hydrogen or a hydrogen-rich gas is passed across the surface of the anode, and oxygen or air is passed across the cathode. With the help of a *catalyst*—a substance that can promote a chemical reaction without changing itself—the hydrogen atoms give up their electrons, thereby becoming hydrogen ions.

In the phosphoric acid fuel cell being developed for utilities, the positively charged ions and negatively charged electrons move in separate paths toward the oxygen at the cathode. In this migration, the hydrogen ions remain in the cell and move through the electrolyte. The electrons, however, move through an external circuit connecting the anode and cathode.

At the cathode, the oxygen, hydrogen ions, and electrons combine to form water, which is the main reaction but not the main point of interest. It is the flow of electrons in the external circuit that is important, since this electron movement actually constitutes electricity.

Fuels

The fuel cells used in the 1960s to power the Gemini and Apollo spacecraft used pure hydrogen and oxygen, but commercial application of this technology will require the use of cheaper fuels. Any fuel that can be converted to a hydrogen-rich fluid is suitable. Naphtha, a light hydrocarbon, is being used in a fuel cell demonstration plant being built in Manhattan, but coal-derived liquids and gases are candidates for the future.

For large-scale power generation, individual fuel cells are assembled into stacks, and the stacks are connected to form power modules. This space-saving arrangement plus the fuel cell's minimal impact on the environment permit fuel cell power plants to be built in areas that have been ruled out for conventional fossil fuel and nuclear plants. Plants can be sited in urban areas that have high electricity demand, cutting losses normally experienced with long transmission lines. Modularity also permits assembly-line mass production, which should lead to reductions in cost and construction time for power plants.

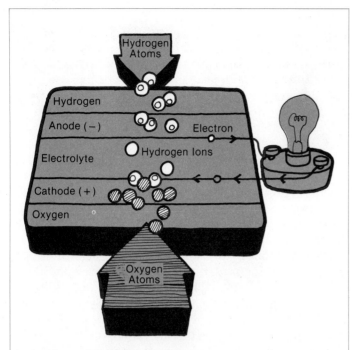

Figure 6-11. Fuel Cell
Hydrogen or hydrogen-rich gas is passed along the surface of the
anode, and oxygen or air is passed across the cathode. The
hydrogen atoms give up their electrons and so become positively
charged hydrogen ions. Meanwhile, the negatively charged free
electrons migrate toward the positive cathode through a connecting
circuit, and this electron flow constitutes electricity.

Unlike conventional generation equipment, fuel
cells are equally efficient at partial and full capacity.
These characteristics make fuel cells almost ideal for
load-following generation.

Costs and Uncertainties

In spite of the many advantages that fuel cells have to
offer, additional efforts will be required before power
plants can be made reliable and inexpensive enough for
commercial use. There is some doubt that suitable fuels
will be available at a low enough cost to make operation
economic.

Because of these uncertainties, large-scale development of fuel cell technology is risky and will require
that the federal government share the extra risk involved
through subsidies. A demonstration plant is now being
built with federal assistance in downtown New York

City. This plant is expected to verify the fuel cell's
advantages in siting, operations, emissions, transmission
savings, and reliability by 1982. If all goes well, commercialization of fuel cells may be a reality by 1985.

MAGNETOHYDRODYNAMICS

With *magnetohydrodynamics* (MHD), the heat energy of
a hot fluid can be converted directly to electric energy.
Since turbines, and the intermediate energy conversions
built into them, have no part in this process, relatively
high efficiencies are theoretically possible.

When electrically conductive material moves across
a magnetic field, an electric current is set up in the
conductor. In a conventional electric generator, wire is
the conductor. An MHD generator uses an electrically
conductive gas instead.

Current Through a Gas

If a gas is heated to a very high temperature, it becomes
ionized: electrons, which carry negative electric charges,
dissociate from the normally charge-balanced atoms,
leaving what remains of the atoms (now ions) with net
positive charges. When this hot ionized gas flows through
the duct positioned in a magnetic field, the field causes
the electrons to move in one direction and the positive
ions to move in the other (Figure 6-12). The movement
of the electrons constitutes an electric current.

Open and Closed Systems

MHD systems can be designed to use the gaseous working fluid in either an open or a closed system. In the
open-cycle system, a fossil fuel—most likely coal—is
burned at temperatures that are high enough to ionize
the gases created in the burning process. The gases are
"seeded" with small amounts of easily ionized potassium
or cesium compounds to increase the working fluid's
electrical conductivity. After the gas has passed through
the MHD electricity generator, it is chemically cleaned
to remove potentially harmful emissions and released.

In the closed-cycle system, heat produced by fossil fuel
burning or nuclear reaction is transferred to an *inert*
(relatively unreactive) gas such as argon, which is also
seeded. The gas is ionized by the heat, passed through
the MHD generator to generate electricity, and then
recycled for reheating, all in one continuous loop. This
system operates at lower temperatures than the open-
cycle system.

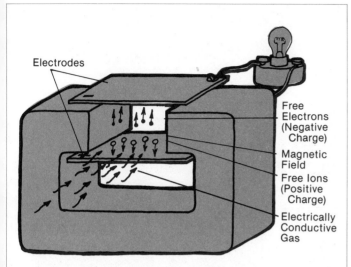

Electrodes

Free
Electrons
(Negative
Charge)

Magnetic
Field

Free Ions
(Positive
Charge)

Electrically
Conductive
Gas

Figure 6-12. Magnetohydrodynamic (MHD) Generator
The gas in an MHD generator is electrically conductive, just as
the metal wire is conductive in an ordinary generator. As this
very hot gas moves through a magnetic field, its positive and
negative charges gather at opposite electrodes. The electrodes are
then joined to allow the negatively charged electrons to flow
toward the positive electrode—in other words, to create an electric
current.

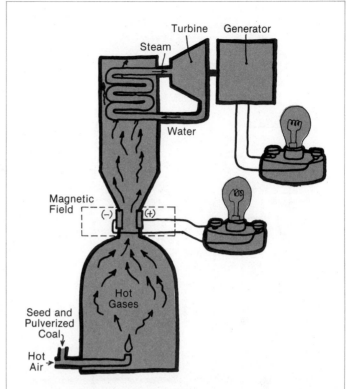

Turbine   Generator

Steam

Water

Magnetic
Field

(−)   (+)

Hot
Gases

Seed and
Pulverized
Coal

Hot
Air

**Figure 6-13. MHD and Steam Turbine Generators
in Combined Cycle**
Hot gases, seeded to increase their conductivity, move through a
magnetic field to generate an electric current. Then the gases,
which are still hot enough to boil water, make steam that spins a
conventional turbine generator to produce additional electricity.

Combined Cycle

Open-cycle MHD electricity generation is receiving
special attention in research circles because it appears
to be suitable for use in a combined cycle with a conventional steam turbine generator. The temperatures
needed to ionize the fossil fuel combustion gases in an
open-cycle MHD process are so high (up to 5000°F) that
even after the working fluid passes through the MHD
generator, it still has enough heat left to boil the water
in a steam turbine boiler (Figure 6-13). Incorporating
MHD generation into a combined cycle could raise the
efficiency of a steam power plant by 15%.

Although MHD power generation seems to offer
considerable potential advantages, not the least of
which is the efficient use of coal resources, development
of a complete large-scale system is still in the early
stages. Areas that need further study include recovery
and reprocessing of seed materials and reliability and
durability of system components. (The very high temperatures corrode the electrodes.) Large-scale commercialization of MHD electricity generation systems
therefore seems unlikely before the year 2000.

BREEDER REACTORS

The breeder reactor is an advanced alternative to the
light water reactor. Like the LWR, the breeder splits
atoms, transforming their nuclear energy into heat that
can drive a turbine generator. In addition to producing
power, however, the breeder is able to produce more
fissionable material than it actually consumes. This
feature gives it a great advantage over the LWR with
respect to economy and full use of fuel resources.

A nuclear chain reaction releases neutrons that breed
(create) more fissionable material when they are absorbed by fertile isotopes such as uranium-238. The
breeder reactor promotes this action by increasing
the average number of neutrons produced during fission

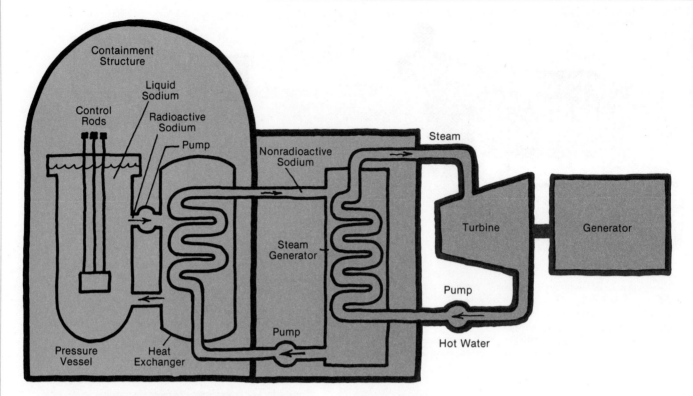

Figure 6-14. Liquid Metal Fast Breeder Reactor (LMFBR)
The fission in the core of the pressure vessel heats the liquid sodium that surrounds the core. This radioactive sodium is pumped into a heat exchanger, where its heat is transferred to the clean sodium in the tube. The nonradioactive sodium then travels through a secondary loop to another heat exchanger, where it heats the water in the tube to produce steam that will turn a turbine generator.

and thereby increasing the number of neutrons available for absorption by fertile nuclei. In addition, since there is no moderator in the breeder to capture neutrons, fewer neutrons are absorbed unproductively.

The Fast Breeder
There are several breeder reactor concepts being studied, but so far the emphasis in both the United States and the rest of the world has been placed on the fast breeder reactor.

The reactor is called *fast* because it uses fast neutrons —that is, neutrons that have not been slowed down by a moderator, as in the LWR. This use of fast neutrons makes some things easier and others harder. It is definitely an asset for breeding because more neutrons are released in a fission reaction caused by a fast neutron than in one caused by a slow neutron. But the probability of fission's occurring at all is much smaller for fast neutrons than for slow ones, which is precisely why a moderator is used to slow down the neutrons in an LWR.

MAINTAINING THE REACTION To maintain a chain reaction, at least one of the neutrons emitted in a fission must go on to cause another fission. If the number of fissions gets so low that a chain reaction cannot continue, the whole process stops, including the breeding. So maintaining the chain reaction must have top priority.

FUEL ENRICHMENT This problem is solved by further enriching, or concentrating, the fissionable material in the fuel rods—to about 12–20% plutonium rather than the 2–3% uranium-235 used in LWRs. Even though fast neutrons miss the target more often than slow neutrons, providing a higher concentration of targets for a given area can ensure a sufficient number of hits. Using greater fuel enrichment in combination with fast neutrons makes it possible to maintain the nuclear chain reaction and still take advantage of the fast neutrons' superior breeding capability.

The LMFBR

The liquid metal fast breeder reactor (LMFBR) is presently the most highly developed version of the fast breeder (Figure 6-14). The liquid metal is sodium, and like the water in an LWR, its function is to cool the core. Unlike water, however, liquid sodium does not moderate (slow down) the neutrons.

The LMFBR has two fuel regions: the *core* and the *blanket.* The fuel mix initially loaded into the core region is generally about 15–20% plutonium-239 (which is fissionable), with the fertile isotope uranium-238 making up the balance. Rods containing this fuel are gathered in hexagonal assemblies, which are arranged to form the reactor core (Figure 6–15). This core is surrounded by a blanket of additional assemblies that contain only uranium-238.

HARVESTING FUEL   Almost all the energy generated by the reactor results from fissions that take place in the core region. The surrounding blanket captures neutrons that escape from the core, thereby creating plutonium from the blanket's fertile uranium-238. Breeding of plutonium also takes place in the core. Together the core and blanket produce more of the fissionable plutonium than was originally consumed in the core. Reprocessing of the fuel is necessary to recover this new fissionable material for future use and to remove unwanted fission by-products.

EXPANDING RESOURCES   The breeding process cannot create more energy than it started with. It just makes more of the energy that is already in the uranium available for producing power.

The considerable nuclear energy in uranium-238 cannot normally be tapped because this material will not fission readily. But by creating fissionable plutonium from this nonfissionable uranium, the breeder can recover most of the available energy from the uranium. Development of the breeder option could expand our usable uranium resources 60 or 70 times. The fuel savings this implies are expected to more than offset the high construction cost of the LMFBR plant.

The breeder has great potential for satisfying the country's growing energy needs: the U.S. stockpile of the

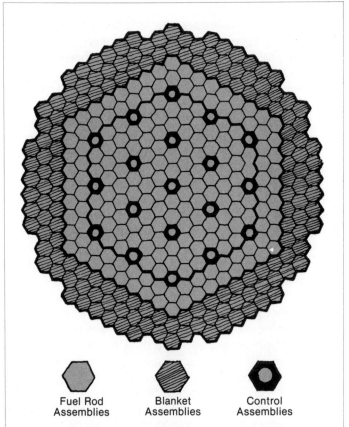

Fuel Rod Assemblies   Blanket Assemblies   Control Assemblies

Figure 6-15. Core and Blanket Cross Section (LMFBR)
The fuel rods in the core of an LMFBR contain both fissionable plutonium and a nonfissionable type of uranium (uranium-238) that actually turns into plutonium when it catches neutrons thrown off by the plutonium during a fission reaction. The blanket is composed entirely of this fertile uranium, and here the breeding of fresh plutonium continues. The harvest of fissionable plutonium from the core and the blanket combined brings in more plutonium than was originally consumed by the power-producing fissions in the core.

fertile isotope uranium-238 that is already mined, refined, and stored is equal in energy content to all U.S. coal reserves, or five times the oil reserves of all the OPEC nations combined. Using the breeder, this fuel supply could be made to last for 1000 years at current rates of consumption without mining another pound of uranium.

Breeder Outlook

There is one LMFBR operating in Great Britain, two in France, and three in the Soviet Union. The United States, where experimental breeders have been producing electricity since 1951, still has no large-scale commercial breeders. Successful operation of the Clinch River demonstration plant, scheduled for completion in about 1990, could lead to the construction of a larger commercial reactor. However, the U.S. breeder reactor development program is now being reevaluated, and commercialization could be deferred indefinitely.

The nuclear fuel cycle is federally regulated. At present, commercial reprocessing and recycling activities are not permitted in the United States, although military reprocessing plants continue to operate. This commercial ban is a major obstacle to the breeder, since the breeder uses reprocessed fuel rather than fuel refined directly from natural uranium ores.

Several European nations are already committed to the breeder option and will no doubt continue to develop the technology as a commercial power source. The future of the breeder reactor in the United States clearly depends on national energy policy. If demonstration and commercialization of the technology were to begin soon, the breeder could be supplying about one quad of energy by the year 2005 and considerably more than that within a relatively short period.

NUCLEAR FUSION

Like nuclear fission, *nuclear fusion* releases some of the energy normally contained in an atomic nucleus by liberating fast-moving neutrons. Whereas energy release in fission involves splitting nuclei, the fusion process releases energy when two nuclei are joined (Figure 6-16). It takes very high temperatures and pressures to push nuclei close enough together for them to fuse. The sun and other stars, with their tremendous heat and gravitational forces, are natural fusion reactors that fuse the nuclei of hydrogen atoms to radiate heat and light.

Hydrogen Fuels

The first controlled fusion reactions produced on earth will probably use an isotope of hydrogen known as tritium. But tritium does not occur in significant quantities in nature. It must be bred from the element lithium, in a manner similar to plutonium breeding in a fission breeder reactor.

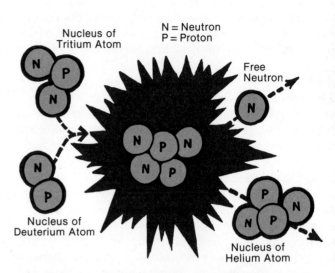

Figure 6-16. Deuterium-Tritium Nuclear Fusion Reaction
Extremely intense heat and pressure can cause the deuterium and the tritium nuclei to fuse, throwing off a helium nucleus and a free neutron in the process. As the free neutron bumps into other nuclei, some of its kinetic energy becomes heat that can be used to make steam for a conventional turbine generator. Nuclear fusion involves no chain reaction, so additional fusions can occur only as long as the proper heat and pressure are maintained.

Another hydrogen isotope that may figure in fusion reactions is deuterium. About one out of every 6700 atoms of hydrogen on earth is a deuterium atom. This sounds like a very small amount, but because of the huge amount of hydrogen available from the water of the oceans, trillions of tons of deuterium are presently available.

Although the principles of fusion are quite simple, the technology is complicated and has not yet been proved efficient for energy production. Fusion research is progressing on a number of approaches, most of which are based on the magnetic confinement of an ionized gas.

Magnetic Confinement

When tritium and deuterium atoms are subjected to the high temperatures required for fusion, they become ionized, separating into negatively charged electrons and positively charged ions. This random mix of free positively and negatively charged atomic material is called a *plasma*.

Probably the biggest technological problem in controlled fusion is finding a way to confine this plasma while subjecting it to extremely high temperatures. Since all known matter vaporizes long before reaching the temperatures needed for fusion, physical confinement will not work. Fortunately, however, because plasma is made up of free positive and negative charges, it will respond to electromagnetic forces, much the way iron filings respond to the lines of force emitted by a magnet. The *tokamak* concept, first developed in the Soviet Union, is the most advanced of several magnetic confinement schemes.

THE TOKAMAK   The tokamak consists of a doughnut-shaped vacuum chamber, called a *torus*, that is surrounded by a complicated system of electromagnets. The magnetic field confines and compresses the plasma inside the torus so that it does not touch the inside walls (Figure 6-17). The compression of the plasma causes it to heat up, and this heat is augmented by a strong electric current or particle beam introduced into the plasma from outside. The torus is lined with a blanket containing molten lithium for transferring heat and breeding more tritium as fuel.

To initiate a fusion reaction, magnetic confinement devices must hold the plasma at a temperature of 180,000,000°F for about one second. So far, the energy required from outside the system to raise this temperature and to power the confinement magnets has been greater than the energy released in the fusion reactions.

REACTOR DRAWBACKS   Achieving a net gain of energy is not the only problem. Designing the device so that the basic structure and many auxiliary systems will withstand the huge magnetic forces is also very difficult. And neutron bombardment of the reactor's wall causes standard construction materials, such as stainless steel, to blister and become brittle, perhaps after only two or three years of operation.

By absorbing neutrons, this wall becomes radioactive, forcing its replacement by remote handling methods. Disposal of radioactive wall and blanket materials from this type of fusion reactor could thus be as problematic as disposal of radioactive waste from fission reactors. However, scientists are testing ceramic, aluminum, and titanium construction materials because these substances do not absorb neutrons readily.

Inertial Confinement
An alternative to magnetic confinement is inertial confinement. Short bursts of laser light or particle beams strike tiny fuel pellets of tritium and deuterium. These

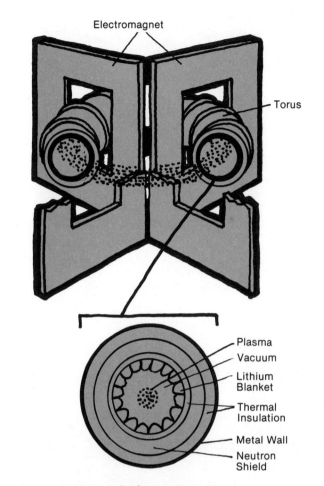

Figure 6-17. Magnetic Confinement Fusion Device With Torus Cross Section
The deuterium-tritium plasma whirls around inside the torus, where it becomes extremely hot because of compression and the charge of an externally generated electric current. To achieve a fusion reaction, the device must keep this plasma at 180,000,000°F for about one second. The lithium blanket transfers heat outward for power generation and at the same time breeds more tritium for use as fuel.

"drivers" heat and compress the pellets in less than a billionth of a second to the point where a fusion reaction occurs.

SMALL EXPLOSIONS   The basic reactor design for laser fusion uses a heavy, spherical pressure vessel with entrance ports in the sides for the laser light (Figure 6-18). Tritium-deuterium pellets arc dropped at intervals of perhaps one second from the top of the vessel with

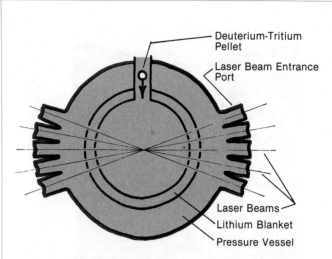

Figure 6-18. Laser Fusion Device
The laser beams strike the deuterium-tritium fuel pellets as they fall into the vessel. Each resulting fusion reaction is a small explosion that releases kinetic energy. When this kinetic energy is converted into heat, it can power a conventional steam turbine generator.

such precise timing that when each falls to the center of the sphere, it is struck simultaneously by as many as 40 high-powered laser beams. Neutrons released by the fusion reaction and carrying most of its energy shoot out toward the vessel walls. Because the laser fusion reaction results in a small explosion within the reactor sphere, the vessel walls must be thick enough to withstand an impact equivalent to that of about 50 pounds of high explosives. The neutrons' kinetic energy is then converted into heat energy for power generation.

INERTIAL VERSUS MAGNETIC   Inertial confinement fusion devices sidestep some of the problems of magnetic confinement devices. They are smaller and lighter and do not require large electromagnets, which are expensive and complicated. On the other hand, they do require extremely large and expensive drivers, which must be aimed, focused, and synchronized very precisely. Even small vibrations in the support structure can cause problems. Drivers must be developed that have tremendous amounts of power and that can be fired once every second—a formidable challenge. In addition, inertial confinement fusion devices face many of the same materials and radiation problems that affect magnetic confinement schemes.

Fusion's Future

Nuclear fusion has the potential for producing great quantities of electricity from cheap and virtually inexhaustible fuel, but not in this century. Fusion development is still in its infancy.

No fusion device has yet produced more energy than is required to run it. Those experimental models that have come close would—if scaled to full size—be larger and far more expensive than would be feasible for commercial use. The first experimental reactors that include equipment for actually generating power are planned for the late 1990s. Even if fusion's problems can be solved, a fusion demonstration plant is not expected to be in operation before 2005, and commercialization of fusion power probably cannot occur before 2010.

THE TECHNOLOGY TIME SCALE

A recurring theme in any discussion of advanced power options is the time it will take for them to make a real contribution to the nation's energy needs. Historically, a new energy process takes 30 to 45 years from discovery to significant commercial use (defined as a 10% share of its potential market). Why does this process take so long?

Development and Integration

The popular image of energy research pictures scientists in a laboratory, working away toward the breakthrough that will bring a new energy technology to the public. Once that breakthrough occurs, the rest is seen as simply a matter of tying up loose ends. But in fact the greatest expenditure of time and money has just begun. As Figure 6-19 shows, a new option must pass through several distinct stages before it can gain a place in the nation's energy lineup.

As a technology passes through these stages, it moves from the simplest performance measure—will it work?—to more difficult issues. What is the best design in terms of cost, safety, and environmental impact? What about reliability and maintenance? Answering these questions requires experiments of increasing scale and complexity, and all of them take time. They also take money, and investment in new technologies is highly sensitive to the many uncertainties that affect energy markets. The result is a long and ponderous process of development and market penetration.

Nuclear power, for example, holds the current record for rapid emergence, and even it took 30 years. Development began in the mid-1940s, commercial entry occurred around 1960, and the nuclear option captured 10% of the electricity market in 1976.

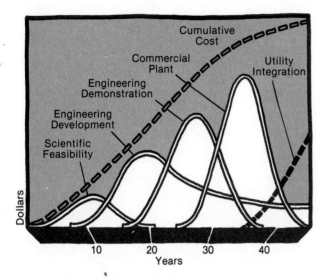

Figure 6-19. Phases of Technology Development and Integration
A new power option passes through several phases before reaching the stage of significant use by utilities. Typically, each of these phases costs more than the one before. Bringing a new power technology to the public, then, requires a steadily increasing financial commitment.

Technology Outlook

The future options now under consideration are unlikely to improve on this record for several reasons. One is that nuclear power, unlike many of today's new options, was economically competitive almost from the start. In addition, today's prospects face growing costs and delays in building the pilot and demonstration plants that must generally precede full-scale commercialization, as well as great uncertainty about the environmental regulations that will affect power production in the future.

The typical technology takes about 25 years to reach the stage of commercial feasibility, and another 20 years to achieve significant use. An unusually complex process such as nuclear fusion may well take the full 45 years for development alone, before the process of market penetration can even begin. This is why work on advanced power options must proceed now, and proceed rapidly, if they are to be ready for the fast-approaching twenty-first century.

Chapter 7

ELECTRICITY STORAGE AND DELIVERY

At 4 o'clock on Sunday morning the demand for electricity is low everywhere—less than half of what it is at 9 o'clock on Monday morning. In fact, power plant operation may dip to only 50% of its capacity at night and on weekends. If baseload energy generated during these off-peak periods can be stored for use during periods of peak demand, power output can be maintained at a more constant level and the generating equipment used more efficiently. Storage could also be the missing link in providing around-the-clock power from intermittent power sources, such as the sun and the wind (Figure 7-1).

HOW STORAGE WORKS

The technology for storing electricity is still, in most cases, highly experimental. Research is under way on a variety of options, including new types of batteries; underground compressed-air storage; thermal energy systems; chemical energy storage, especially hydrogen; superconducting magnets; and flywheels. We will also look at pumped-hydro storage, the most common method of energy storage used today.

Pumped Hydro

Pumped-hydro storage, which has been in use in the United States since the 1930s, is not direct storage of electricity. Electricity is converted to potential energy, which is then converted to kinetic energy and finally back to electricity. Baseload generators running during periods of low demand produce power to pump water from a low reservoir into a higher one. Later, the water

is allowed to flow down from the higher reservoir through turbines, generating hydroelectricity to meet demand peaks (Figure 7-2).

Underground pumped hydro is still in the development stage, but it is similar to the conventional method, except that one or both of the reservoirs are located underground. The principle is the same, with one reservoir higher than the other. A large percentage of the energy storage used by electric utilities in the next 25 years will probably come from conventional and underground pumped hydro.

Batteries

The oldest and most familiar method of storing electric energy is the battery. The battery works by transforming the energy of a chemical reaction directly into electricity. The kind of battery used in a flashlight is called a primary battery. It is used until the chemical reaction is exhausted, and it cannot be recharged. The battery that starts a car motor is a secondary, or storage, battery and is rechargeable by reversing the chemical reaction in the battery.

The main problem with batteries is the relatively small amount of electricity that can be stored in relation to cost, weight, and size. Of the rechargeable batteries that have been developed, none is inexpensive enough to be practical for utility storage use. Two promising new storage batteries that are being developed for large-scale utility use are the sodium-sulfur and zinc-halogen batteries. When developed, these types should be less expensive than today's batteries.

Such advanced battery systems are expected to reach the commercial market by the 1990s. Batteries have

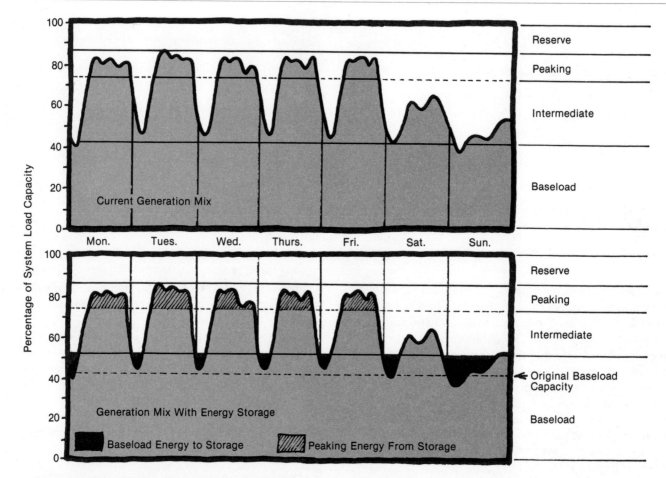

Figure 7-1. Impact of Energy Storage on Power Systems
If large-scale energy storage were available, utilities could use economic baseload generators to charge the storage system during off-peak hours. Discharge of the stored energy during periods of peak demand would then reduce or replace the need for expensive-to-run peaking generators. Besides raising the efficiency and holding down the cost of the generating system, energy storage could help conserve the oil and gas that would otherwise be used to fuel these peaking generators.

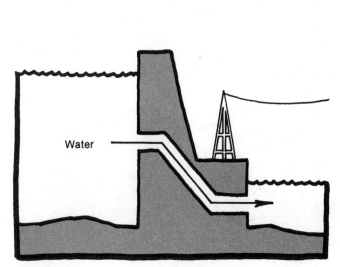

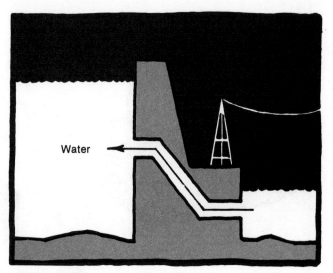

Peak Hours

Off-Peak Hours

Figure 7-2. Pumped-Hydro Storage
During periods of peak demand, water flows through the turbine to a lower reservoir, producing hydroelectric power in the process (left). During off-peak periods, inexpensive power pumps the water back into the upper reservoir, where it waits to be used again (right). The United States has used pumped-hydro energy storage since the 1930s.

minimal effects on the environment and are especially attractive because they could be dispersed throughout the utility system, allowing electricity to be stored close to where it will be needed. Figure 7-3 shows how batteries could be used for storage of off-peak energy.

Compressed Air

Compressed-air storage is a demonstrated technique (a plant has been operating in Huntorf, West Germany, since 1978) for reducing the oil or gas needs of gas turbine power plants. Off-peak power is used to compress air, which is stored in an underground reservoir for later use, when it is combined with fuel combustion to drive the gas turbines during peak demand. The compressed air allows a gas turbine to produce a kilowatt-hour of energy from only 4000 Btu of oil or gas instead of the 12,000 Btu required by conventional gas turbines. Figure 7-4 shows how such a system would work.

Thermal Storage

Thermal storage is simply the storage of energy as heat. This method is not cost-competitive with other techniques, such as pumped hydro, for use in power generation, but it can be used in the home. For example, off-peak electric power can heat water to be used throughout the day. For space heating, ceramic or iron storage elements in a heater or furnace that are heated during the night can retain enough warmth for use all day long.

Chemical Conversion

Chemical energy conversion systems other than batteries offer another possibility for energy storage. The trick is to convert the fuel from a low-energy chemical state to a high-energy state by applying heat or another energy form (such as electricity) and then return it to the

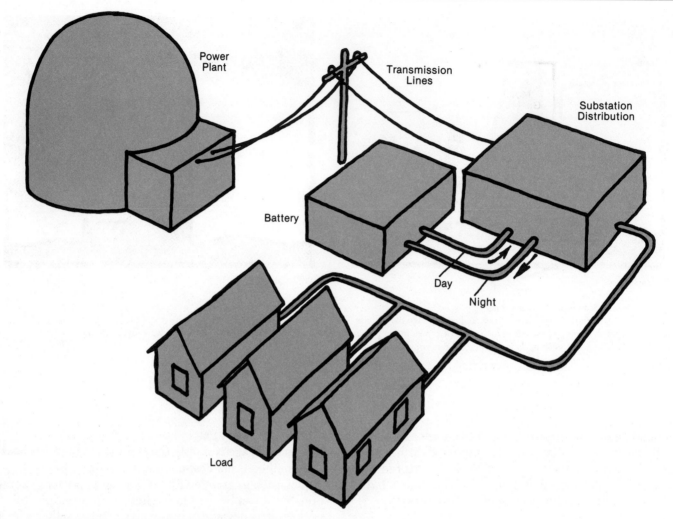

Figure 7-3. Battery Storage in a Typical Utility System
Batteries that can store large amounts of electricity relative to their weight and size will be a welcome addition to utility networks. As in other storage schemes, the batteries would be charged during periods of low demand and discharged during periods of high demand. Battery storage is expected to become available for large-scale utility use by the 1990s.

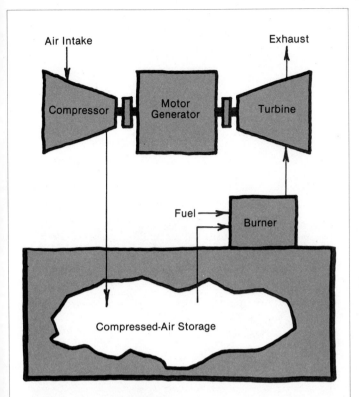

Figure 7-4. Compressed-Air Storage
Large reservoirs, such as underground caverns, are key to the use
of compressed-air energy. Off-peak power drives the motor to
compress the air for storage. When it is later released, the
expanding air can be used along with fuel combustion to drive a
gas turbine for peaking power.

original state, with an accompanying release of usable
energy. The production of burnable hydrogen from
water is the most familiar example of a chemical energy
conversion system. When electricity is applied to ordinary water, the water decomposes into hydrogen and
oxygen—a process called *electrolysis,* which is simple
enough in principle to be a common demonstration in
high school science classes.

Hydrogen has many advantages as a fuel. When it
burns in ordinary air, the only emission besides water is
a low level of nitrogen oxides. If combustion takes place
in oxygen alone, water is the only by-product. Hydrogen

could be produced by electrolysis during off-peak hours
and stored in dispersed locations, close to where it
would be used to supplement gas or to generate electricity at periods of peak demand (Figure 7-5).

Like natural gas and oil, hydrogen can travel in
pipelines. It can be stored as a high-pressure gas or as a
very low-temperature liquid. It can also be absorbed in
such metals as magnesium, lithium, or iron-titanium
alloys to produce a metal hydride. The hydrogen is then
easily released by heating the metal hydride.

Although hydrogen is an extremely versatile fuel, as
an energy source it has one major drawback: costs are
presently much higher than for other storage systems.
The same is generally true of other chemicals that can
be produced as storage media.

Superconducting Magnets
Magnetic energy storage is the only storage system in
which electric energy is stored directly. An *electromagnet* is an iron cylinder or doughnut shape, wound
with conducting wires that carry electricity. Under
ordinary circumstances the windings must be continually
supplied with energy because of losses that result from
electrical resistance. But superconductivity means that
at very low temperatures—far below zero on the familiar
Fahrenheit scale—certain conducting materials lose
nearly all electrical resistance to direct currents. Once
a direct current is introduced at such temperatures, it
will persist indefinitely and the magnetic field will be
maintained for future energy withdrawal. Obstacles to
the use of superconducting magnets are size and cost.

Flywheels
A *flywheel* is a wheel that can absorb, store, and give up
energy as required. Energy is stored when the wheel is
set to rotating. When energy is withdrawn from the shaft
as work or is dissipated as friction, the wheel's rotation
slows down. Flywheels are commonly used to smooth the
pulsed power derived from engines. In the electric
utility industry, energy stored in giant flywheels during
off-peak hours would be used to drive generators at
times of greater demand.

Although flywheels have been used for centuries—as
the basis of the potter's wheel, for example—they have

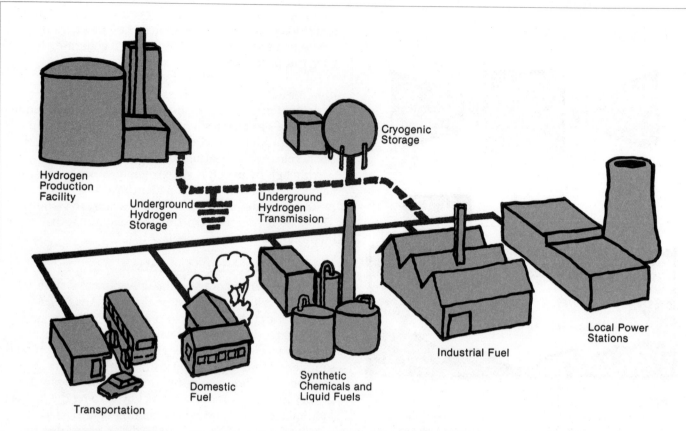

Figure 7-5. Hydrogen Storage
Off-peak electricity could be used to decompose ordinary water into oxygen and hydrogen. The hydrogen, a clean and highly combustible fuel, could then be stored in dispersed locations, close to where it would be used directly or converted back into electric energy for periods of peak demand.

not been practical for energy storage because of the large size and high speed that would be required and the cost of materials that could withstand the resulting centrifugal forces.

Prospects for Energy Storage

Energy researchers now estimate that up to 5% of the nation's total electric energy needs and up to 17% of peak-time electricity can eventually be supplied from energy storage systems.

Until at least 1985 pumped-hydro and compressed-air storage will be the only practical storage methods. Advanced battery systems may become commercially feasible after 1990. But other storage possibilities, such as hydrogen, flywheels, and superconducting magnets, appear less promising unless major technological break-throughs can reduce their costs.

It is important to remember that energy storage is not an energy source. It is simply a method of smoothing the utility load pattern, so that the more efficient and cheaply fueled baseload plants can operate around the clock, reducing or eliminating the need to rely on older and/or less efficient plants.

THE T&D SYSTEM

Although the role of storage is growing, most electricity is still used as soon as it is generated. This means moving electricity in bulk from one place to another through a system of transmission and distribution (T&D) lines. Electricity cannot be loaded onto a train or a truck or transported across the country in a pipeline. A metallic

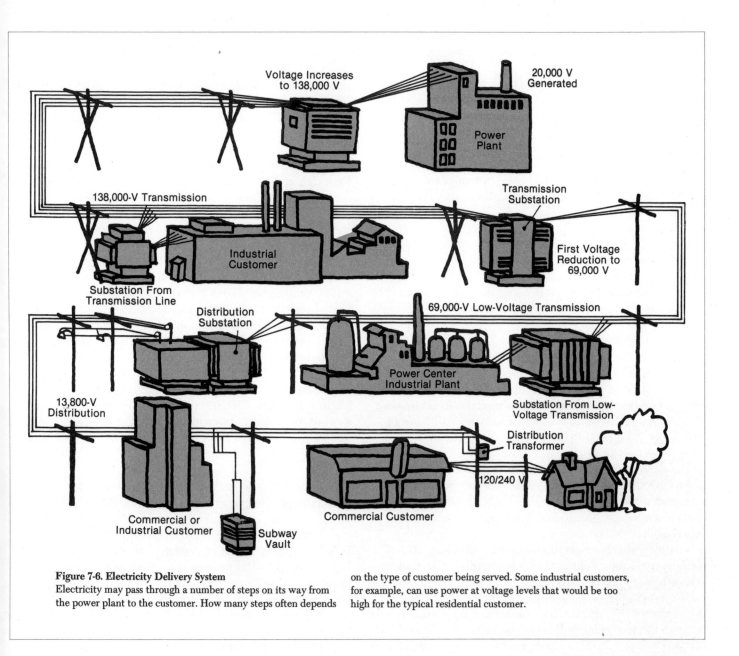

Figure 7-6. Electricity Delivery System
Electricity may pass through a number of steps on its way from the power plant to the customer. How many steps often depends on the type of customer being served. Some industrial customers, for example, can use power at voltage levels that would be too high for the typical residential customer.

connection—wires—must exist every inch of the way from the generator to the lamp on your desk.

Figure 7-6 shows the steps in the electric supply system from power plant to customer. The *transmission* system transports bulk electric energy from the generator to the main substation that serves a given area. There its voltage is reduced, and it is passed on to the distribution substation. From that point, *distribution* lines carry the electricity into cities and neighborhoods and into the homes, offices, and factories of individual customers.

This whole interlocking system of electricity delivery is commonly known as the *grid*. About 20% of the grid consists of transmission lines. The remaining 80% consists of distribution lines.

In addition to providing point-to-point connections between generator and distribution substations, transmission lines serve two other main functions. The first of these is the pooling of energy reserves between large power systems. Individual electric utilities use transmission lines to interconnect their systems for pooled operation. This interchange provides more reliable service to the customer and allows the power plants to operate more economically.

Second, and especially important in times of threatened power shortages, is that a utility having trouble meeting peak demand due to fuel cutoffs or plant shutdowns can actually buy power from other utilities and have it delivered through the transmission system. This allocation system headed off serious power outages in the mid-Atlantic states when coal-fired power plants could not get the fuel they needed during the coal strike in the winter of 1977.

The demand for electricity is growing, but simply generating more will not meet that demand. It does not matter how many new power plants are built unless the electricity they generate can be delivered to the customer. This means that T&D capability must increase to match the increase in power generation.

More Power, Less Space

Large amounts of land are required to move electricity. Just one mile of overhead transmission line can take up 15 to 20 acres of land for its long, narrow corridor. Building transmission lines over railroad tracks is one way to promote efficient land use. But utilities still have to find ways of increasing the amount of power that can be transmitted through a line while decreasing the width of the corridor, or *right-of-way*. Two possible solutions are to use ultrahigh voltage (UHV) or to construct compact power lines. Figure 7-7 shows how higher-voltage transmission can cut down on the amount of land needed for rights-of-way.

In the 1960s, lines rated at 345 and 500 kilovolts (kV) were introduced. Today, the highest-voltage lines in operation are rated at 765 kV. If the growth in transmission capacity is to keep pace with the growth in generating capacity, the transmission of bulk energy will require the use of UHV lines in the range of 1000–1500 kV. Experimental UHV lines at 1200 kV have already proven feasible. Further, the cost of transmission decreases as the voltage increases, at least up to the 1200-kV level.

The main obstacles to UHV installation are environmental issues, including audible noise, interference with radio and television receivers, and the possible effects of electric fields on biological systems (see Chapter 8).

Each of the parts of the immense power networks is the subject of constant research and development. Here are some of the ways that this work is improving the old T&D system and preparing it for the growing demands that lie ahead.

Poles and Towers

The most familiar features of the T&D system are the poles and towers that support the overhead wires. Most poles now in use are made of wood, so utilities must inspect and replace them regularly to keep them safe from groundline decay.

Researchers are working on new chemical fumigants to make wooden poles last longer. The chemicals, poured into holes drilled in the pole near the groundline, vaporize and travel deep into the wood fibers. Pole decay can now be arrested for seven years or more, and it is hoped that regular doses can make poles last indefinitely.

Utilities are also experimenting with power poles made from composite wood material. This process uses chips or flakes from wood that has no other commercial value, such as scrub oak and red alder. These chips are bound together with phenolic resin, an adhesive of great strength, and pressed into long sheets similar to very highly compressed particle board. This board is then laminated into power poles that are both cheaper and stronger than natural wood.

The towers that support long spans of overhead transmission lines are usually made of metal—galvanized steel or aluminum. They are then latticed, braced, and guyed in a variety of ways to ensure strength and stability.

Wires

Conductors, the wires through which electricity flows, are made of several different materials. Most high-voltage transmission lines today use ACSR (aluminum conductors, steel-reinforced). Although its conductivity is only about 60% that of copper, the lower price and lighter weight of aluminum give it an advantage for most uses. Aluminum alone is not strong enough to be used in longer spans, however. So aluminum conductors are wrapped around steel cables to provide the necessary strength.

Sometimes a conductor will discharge electricity into the atmosphere, resulting in a halolike glow called a *corona*. This corona does not present a health or environmental hazard, except for its audible noise, but it does represent a loss of power. The severity of this loss can depend on weather conditions—on whether it's raining, for example—and whether there is dirt on the wires. Reducing or eliminating corona effects is one research target for more efficient power delivery.

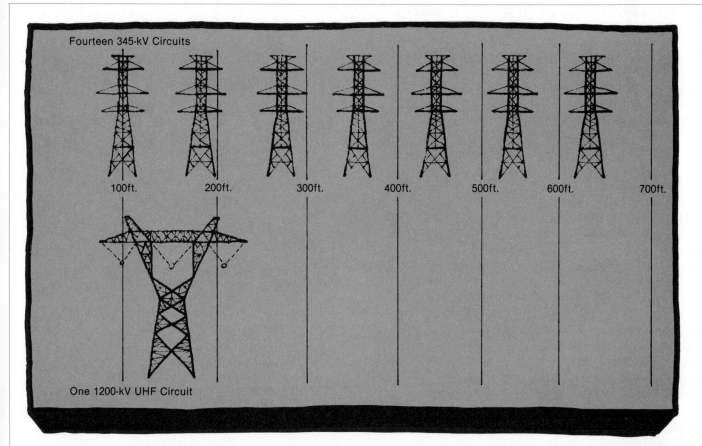

Figure 7-7. Ultrahigh-Voltage Transmission and Land Use
Comparing corridor widths shows why power transmission is more efficient, from the standpoint of land use, at higher voltages. The two corridors shown can carry identical amounts of electricity. The single 1200-kV line in the bottom corridor requires a higher tower, but it packs the same power into a right-of-way that is much narrower than the one above.

Insulators

On overhead lines, the insulation between the conducting wires and between the wires and other objects is provided by air—by the space between them. At the towers or poles, utilities use insulators made of glass or porcelain, substances that conduct very little electricity. Insulators come in many types, sizes, and shapes, but they all have one job to do: to keep electricity from flowing where it is not wanted—that is, any place except through the conductor.

The efficiency of the insulating material can have a substantial effect on the cost and reliability of conductors and of substation equipment. For example, porcelain alone often accounts for 25% to 30% of these costs. Current research is pursuing a cheaper replacement for porcelain or ways of improving its insulating properties without increasing the cost of manufacture.

Hazards

VIBRATION   Movement of high-voltage lines created by wind and ice can cause severe problems. Some of these vibrations can be prevented or lessened by the use of *dampers*, metal weights attached to the lines at points calculated to break up the motion.

A more extreme type of vibration is known as *galloping*. The nonrhythmic swaying of the wires may become so great that they whip together, causing the circuit to fail. However, a recently developed device called a *galloping detuner* shows promise for resolving this problem. The purpose of the device, which is fairly inexpensive and may be attached to new or existing lines, is to stop the motion of the line before it gets started, rather than absorbing the motion of a moving line, as the damper does.

LIGHTNING   Lightning is another natural phenomenon that poses a hazard. When lightning strikes an overhead line, it creates voltages that are much higher than what the line normally carries. This surge of energy can be destructive to the line itself, as well as to other parts of the system. To prevent this damage, lightning arresters are designed to bleed off the high-voltage surge, without allowing the normal flow of power to escape as well.

FOREIGN OBJECTS   High-voltage surges can also stem from the brief short circuits caused by tree limbs, kites, or other foreign objects falling or blowing against the line. In fact, the cost of trimming trees to keep them clear of overhead lines is a sizable item in utility companies' annual maintenance budgets. So researchers are now exploring methods of controlling tree growth that will not harm the health or the appearance of the tree.

Substations

Substations are the gateway for the transfer of all power from generator to consumer. Substations at the power plant itself first transform the power generated to higher voltages for transmission, but the main function is then for each local substation to reduce the voltage to a value appropriate for local distribution and send the energy on its way to the consumer (Figure 7-8). The substation is also a convenient place for the overhead portion of a transmission line to be connected with the underground portion.

The *transformer* is the key component of the substation, accounting for a major share of the capital costs of the transmission system. A *step-down* transformer is a device for changing the comparatively small current at high voltage that is transmitted from the power plant to the local substation into a larger current of lower voltage,

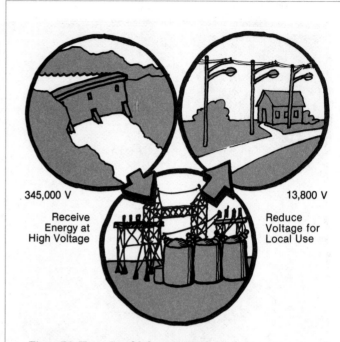

345,000 V
Receive Energy at High Voltage

13,800 V
Reduce Voltage for Local Use

Figure 7-8. How a Local Substation Works
The substation is the gateway for the transfer of all power from the generating plant to the customer. The substation's transformer steps down the voltage of the power that it receives from the generator to make that power suitable for use by homes and businesses.

suitable for distribution to consumers. A local power station transformer may be as big as a small house, and it is usually full of oil that circulates to carry off the excess heat.

Some substations are simply switching stations where different connections can be made between various transmission lines, while others are a complicated maze of step-down transformers, high-voltage switches, insulators, circuit breakers, and lightning arresters, as well as relaying and metering facilities. Since most substations are located in or near the populated areas they serve, their esthetic and environmental aspects are particularly important. Reliability is also crucial to uninterrupted customer service, and a great deal of research is being directed toward the development of equipment that will allow substations of all sizes to handle the increasing demands that are being placed on them.

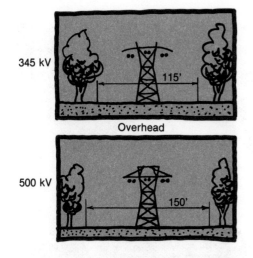
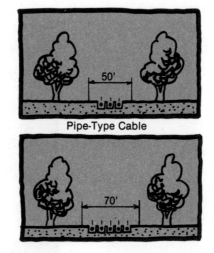
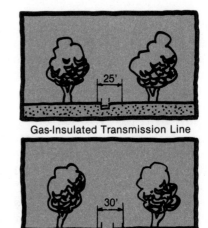

345 kV — Overhead — 115'

500 kV — 150'

Pipe-Type Cable — 50' — 70'

Gas-Insulated Transmission Line — 25' — 30'

Figure 7-9. Overhead and Underground Transmission Corridors
Overhead transmission lines (left) typically require a wider corridor than underground lines (center and right). Overhead transmission is still from 5 to 15 times cheaper, though, so new transmission lines are placed underground only where there is no alternative. In the case of local distribution lines, the cost difference between overhead and underground lines is not so extreme, and many local lines are placed underground for reasons of space and appearance.

UNDERGROUND VERSUS OVERHEAD

There are only two places that power lines can be located: overhead or underground. Most of our current network is overhead. The reason is that overhead power transfer is more efficient and also cheaper.

Most new distribution lines are being installed underground because overhead space is becoming very scarce in urban areas and because people have grown increasingly sensitive to the appearance of overhead wires and poles. The power corridor underground is not only narrower, it is—most important from an esthetic standpoint—hidden from view (Figure 7-9).

Almost 300,000 miles of primary (5–15 kV) underground cable have been installed in the United States in the last 15–20 years. Many miles of secondary (120–240 V) cable have also been placed underground. The rate of primary cable placement has now grown to about 20,000 miles per year. Because underground distribution lines are so comparatively expensive, however, no effort is being made to replace existing overhead lines with underground systems.

The cost of underground distribution is from 1.5 to 5 times as high as its overhead counterpart, while the cost of underground transmission is from 5 to 15 times as high. For this reason, new transmission cables are placed underground only where there is little alternative. It has been estimated that if only *new* T&D lines were placed underground, the cost to the utilities—and therefore to the consumer—would be around $15–20 billion per year. It is unlikely that underground systems can ever be installed at anywhere near a competitive cost.

Installation

The way in which an underground cable is installed or replaced depends on the type of cable, the voltage it will carry, and the nature of the terrain. In open fields, or under rural or suburban roads, the cable may be laid directly in the ground. In cities and suburbs, it is often installed in pipes or drawn through ducts for protection.

Installation and construction can account for as much as 50% of total underground cable costs. And the procedure of digging a trench, laying the cable, and then refilling the trench is awkward and disruptive as well as very expensive.

One improvement now being tested and evaluated is a means of cutting concrete by using water jets—a method that promises to be faster, cleaner, and quieter than current techniques, which use jackhammers or concrete-cutting saws. Research is also under way on a small, accurately guided boring device called a *mole,* which would follow the line of an underground cable without the necessity of digging an open trench. Another research project uses a radarlike system to detect and identify underground obstacles. The ability to plot the best cable route before digging begins should result in substantial savings.

Repair and Maintenance

Although overhead lines are prey to the destructive forces of wind, ice, lightning, and vandalism, faults are fairly easy to find and repair. Underground lines are free from most of these hazards. But when a fault does occur in a buried cable, locating and repairing the problem is likely to be a difficult, expensive, and time-consuming task.

The total number of faults occurring on underground T&D cables in the United States is estimated at around 165,000 each year. About 90% of these faults are easily found and repaired, while the remaining 10% account for the major share of the costs of fault location and repair. These costs are very high, and they are expected to grow rapidly—to as much as $128 million annually by the year 2000—unless there is a significant advance in technology.

Cable Performance

Most research focuses on improving the efficiency of the underground cable itself. And here the biggest problems are overheating and corrosion.

When carrying current, underground cables lose energy that escapes as heat. There is no place for this heat to go except into the soil. If even a small amount of heat is released in excess of the soil's ability to absorb it, it can build up to dangerously high temperatures around the cable. The overheated cable insulation may then deteriorate, causing a short circuit.

Underground cables need very heavy insulation. And the higher the voltage, the more insulation required. The most common type for high-voltage lines is paper tape insulation combined with oils. This oil-impregnated paper is wrapped around one or more conductors and enclosed in a protective sheath (Figure 7-10).

The corrosion of conductors—usually concentrically wound copper—is a matter of increasing concern. Cable life in some areas has been cut from an expected 35–40 years to only about 5 years. Recent research shows that it is possible to eliminate this problem on new construction by applying a special jacket over the wires to protect them from corrosion. If premature replacement of only 1% of the 20,000 miles of underground distribution cable now being installed each year could be avoided, the savings could be as high as $2.6 million annually.

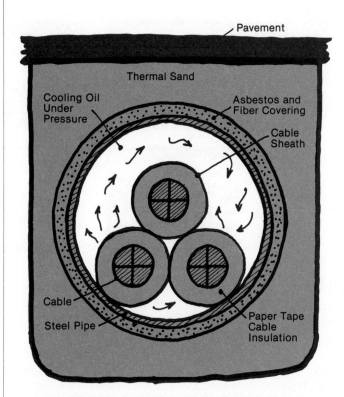

Figure 7-10. Cross Section of Typical Underground Oil-Filled Cable
Because cables need heavy insulation, they are subject to overheating. The higher the voltage, the more insulation is required. Circulating oil helps carry off heat from the heavily taped cables.

DIRECT CURRENT TRANSMISSION

Although direct current (dc) transmission has seen little use in the United States, the concept has been around for a long time. In the 1880s and 1890s, there was widespread controversy over whether direct current or alternating current (ac) systems would be best for meeting the nation's electricity needs.

A Checkered History

Edison and his backers were on the side of tried, safe dc. The problem was that dc generators could produce power only at lower voltages. Further, once dc was generated, there was no useful way to transform (increase or decrease) its voltage. So dc power could be sent out only at low voltages, which made it costly and inefficient to deliver power to distant points. Edison's dc plan would have meant locating power plants very near the consumers they were to serve.

Westinghouse and his backers favored ac. The main advantage of ac was that its voltage could be stepped up for low-cost transmission and then stepped down again for low-voltage consumption. This meant that an ac power station could have a far greater service radius than a dc station. Transmission of ac was more efficient than that of dc simply because it could use higher voltages and so deliver more power per mile of transmission line.

There was doubt, however, about whether ac could be used safely for commercial purposes. There were difficulties in keeping ac voltage and current under control. It was not until scientists found a way to make ac manageable and effective commercially that ac triumphed, and dc fell into a long period of dormancy.

Direct Current Reemerges

Direct current technology began to emerge again only around the time of World War II. In fact, dc had always had certain advantages over ac. Being inherently stable, dc required smaller lines, less insulation, and narrower rights-of-way at any power level. But one fact prevented the widespread use of dc transmission: the converting equipment needed to make it economically feasible was simply not available.

Because dc systems could not readily transform voltages—raise them for transmission from the power plant and then lower them again for the consumer—most electricity was both generated and consumed in ac form. The problem was to find a means to convert this ac power to dc for travel and then convert it back to ac for use. That way we could benefit from the best of both systems: the stepped-up, higher voltages possible with ac could be carried along dc's narrower and less equipment-packed corridors.

Rapid technological advances since World War II, and especially since the 1960s, have now broken the barriers to ac-dc conversion. High-voltage direct-current (HVDC) transmission has become a reality both in the United States and abroad. Given the advantages of dc transmission, it is expected to grow substantially as a complement to existing ac power systems (Figure 7-11).

The Promise of HVDC

Where scarcity of land and concern with the appearance of power equipment favor the use of underground cables to feed electricity into urban areas, HVDC may be the system of choice. Underground ac systems face increasing problems with short circuits and voltage control as the length and load of their transmission lines increase, but dc does not have these problems.

The national drive toward energy self-sufficiency

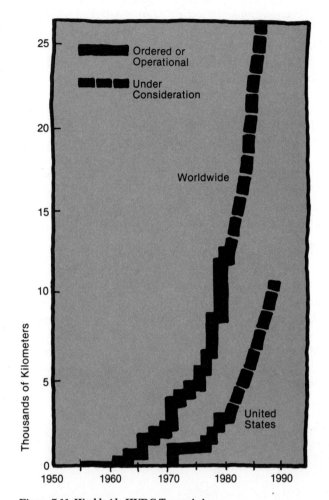

Figure 7-11. Worldwide HVDC Transmission
The transmission of direct current is expected to grow substantially in the next five years as a complement to existing ac power systems. European nations, especially Sweden, have pioneered this development, and HVDC systems now exist in Asia and Africa as well. Underwater lines, such as the line linking Great Britain and France beneath the English Channel, are an especially important application for dc power transmission.

could also mean greater dc use. Opening up western coal fields far from urban areas could mean the location of new coal-fired power plants far from those who consume the power. The possible location of nuclear power plants in remote areas would also push the need for long-distance HVDC transmission.

Another important role for dc lines is to connect existing ac systems. In many places it is virtually impossible to link neighboring ac systems because uncontrollable power instabilities could result. In some places such connections are feasible but would not be stable under conditions leading to cascading blackouts.

Links of dc can act as buffers. They can control the power flow by converting ac to dc and then back to ac again. The dc link essentially decouples the two ac systems, removing the need to attempt synchronization of their frequencies.

These benefits, plus the fact that dc transmission cuts in-transit power losses, suggest a bright future for HVDC. Costs have also come down in recent years, and experts predict that this trend will continue for at least another decade. Meanwhile, the potential applications of dc transmission continue to grow as converter reliability increases.

A New Partnership

Unlike the situation almost 100 years ago, ac and dc are no longer rivals. They are now partners in the business of delivering power from its point of production to its point of use. Blending the talents of ac and dc can mean new flexibility in meeting the power delivery needs of vast service areas with many variations in geography and population patterns.

There is little prospect, however, that dc will become the dominant partner. Most of our electrical appliances and business machines, from the home air conditioner to the corporate computer, are designed to run on ac power. This represents a huge consumer investment, and one not readily abandoned.

Further, dc power still cannot be stepped down to a level appropriate for home and business use. No one needs 250,000 V in a living room or office to run 120-V appliances. The ability of ac power to be transformed from very high to very low voltages remains an essential one in our system of power delivery.

Growing Delivery Costs

Power delivery systems are expensive. The huge network of T&D lines, transformers, switches, and protective equipment extending from power plant to consumer is valued at nearly $60 billion. The T&D system accounts for about 30% of the total cost of providing power.

The need to move more and more power over greater and greater distances will mean expensive new equipment carrying higher voltages. In addition, increasing sensitivity to appearance and to environmental issues will probably hasten underground cable installation, although the result could multiply our current costs for power delivery. Whatever the direction of T&D system expansion, it is certain that it must expand rapidly to keep pace with the growing demand for power.

Chapter 8

ENVIRONMENTAL CONCERNS

What is harmful or undesirable to human beings may be natural or made by human activity. Nature, for example, regularly contributes dust to the air and sediment and bacteria to water. There is really no "clean air" or "pure water" in nature.

People add to nature's load of undesirable substances by injecting waste products into the environment. Nature can absorb a certain amount and variety without disruption, but when we inject more waste products—in the form of gases, heat, toxic chemicals, sewage, and garbage —than the environment can absorb, we upset the balance of natural systems.

Industrialized nations have been consuming natural resources and producing waste products at an unprecedented pace, especially since World War II. We now realize that we have sometimes exceeded nature's ability to handle these wastes in a positive or beneficial way. So we must now look for a new balance that will allow us to enjoy the benefits of industrial civilization without overloading nature's absorptive capacity.

BALANCING BENEFITS AND RISKS

We take risks every day in exchange for convenience. For example, if we drive a car, we could be involved in an accident. Figures indicate that the risk is high for an average driver, but motorists apparently find such risk acceptable. Many people do not drive cars, but virtually everyone uses electricity. As a society, we are even less likely to give up our dependence on electricity than on the automobile. Fortunately, the risks associated with everyday power production are lower than those of everyday auto use. Yet there is no escaping the fact that

electricity generation, like any other complex human endeavor, is bound to have some impact on people and their surroundings.

About 80% of the total pollutants emitted into the atmosphere are natural. Only about 20% result from human activity. However, because human-made pollutants are generated largely in urban and industrial areas, these areas bear most of the impact, and it is here that control efforts focus (Figure 8-1).

AIR QUALITY

Coal combustion creates by-products. Among the most important of these by-products are sulfur oxides, nitrogen oxides, particulates, and carbon dioxide.

Sulfur Oxides

Sulfur is an element that is present in almost all coal, although some kinds of coal contain more sulfur than others. During combustion, the sulfur combines with the oxygen in the air to form sulfur dioxide. Fuel combustion, mostly of coal, accounts for about two-thirds of the sulfur dioxide released by human activity in the United States every year.

As the sulfur dioxide mixes further with oxygen and trace substances in the air, a variety of sulfate compounds emerge. How these transformations take place, and in what proportions, is a subject of vigorous research, but the process remains somewhat mysterious. The behavior of sulfur emissions depends partly on the type of coal used and how it is burned. In addition, the presence of light, moisture, and other pollutants in the atmosphere may also be important in triggering the complex changes that sulfur emissions undergo.

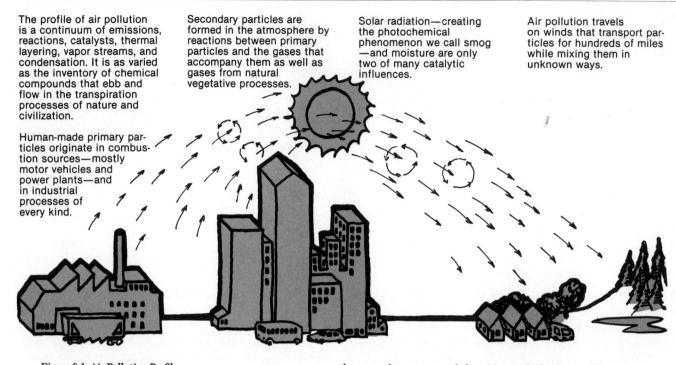

The profile of air pollution is a continuum of emissions, reactions, catalysts, thermal layering, vapor streams, and condensation. It is as varied as the inventory of chemical compounds that ebb and flow in the transpiration processes of nature and civilization.

Human-made primary particles originate in combustion sources—mostly motor vehicles and power plants—and in industrial processes of every kind.

Secondary particles are formed in the atmosphere by reactions between primary particles and the gases that accompany them as well as gases from natural vegetative processes.

Solar radiation—creating the photochemical phenomenon we call smog —and moisture are only two of many catalytic influences.

Air pollution travels on winds that transport particles for hundreds of miles while mixing them in unknown ways.

Figure 8-1. Air Pollution Profile
Only about 20% of the total pollutants in the world's atmosphere result from human activity, but this percentage is concentrated in the areas where most people live. Motor vehicles, heavy industry, and power plants all contribute, and chemical reactions in the air compound the problem.

HEALTH AND ENVIRONMENTAL EFFECTS

From the days of the smothering London fogs of the 1800s, the sulfur in coal has been blamed for the discomfort and illness people experience during episodes of acute air pollution. But recent research indicates that sulfur dioxide by itself, at commonly existing levels, probably is not a serious health hazard. Rather, its presence in the air tends to indicate the presence of other pollutants, some of them derived from the sulfur dioxide by chemical reactions. These other pollutants, some sulfur compounds and some not, are the ones that most likely bear watching.

Every city dweller is familiar with the symptoms of air pollution: red eyes, scratchy throat, perhaps a headache or a cough. It is known that extremely high levels of air pollution can trigger a temporary increase in the death rate, especially among the elderly and the chronically ill. What is not known is the long-term effects (if any) of lower, everyday pollution levels on people who are otherwise healthy—that is, on the majority of the population.

It is often impossible to separate the effects of different pollutants because they all occur at the same time. Moreover, it is difficult to tell the health effects of air pollution from those of other stresses, such as winter cold and summer heat, tobacco smoking, and poor nutrition. Pinpointing exact cause-effect relationships between specific air pollutants and specific illnesses requires very complex studies, and the results so far are not conclusive.

One important effect of sulfur emissions is reduced visibility. There is no longer much doubt that sulfur compounds contribute heavily to the murkiness typical of air pollution episodes.

Sulfur emissions are also suspected of playing a part

in the phenomenon known as *acidic precipitation*. The theory is that sulfur in the atmosphere, some natural and some human-made, combines with oxygen and water vapor to form dilute solutions of sulfuric acid that precipitate as rain. Scandinavia has experienced acidic precipitation for decades, and more recently people have begun to recognize it in various places in the United States. Its effects are disputed; some people contend that acidic precipitation can kill fish in lakes, impair forest productivity, and damage commercial crops.

Further research is clearly needed to find out what sulfur emissions really do, alone or in combination with other substances. The results will help in making better decisions about what substances need controlling, and to what degree.

CONTROL TECHNOLOGY   Today's control of sulfur emissions from power plants is primarily by use of stack gas scrubbers. Stack gas *scrubbing* is a process in which the gases that result from coal combustion are passed through tanks containing a material, such as lime or limestone in a water slurry, that captures and neutralizes the sulfur dioxide. The sulfurous gases cannot escape into the atmosphere because they react with the lime or limestone solution to form a thick sludge. Recently, dry scrubbers have been built that capture sulfur dioxide using a dry powder and a fabric filter. Scrubbers clean between 80% and 90% of the sulfur dioxide from flue gases, but the cost is high.

Nitrogen Oxides

Nitrogen is a colorless, odorless gas that makes up about 78% of the atmosphere. Nitrogen combines with oxygen and water to form several nitrogen oxides. The most important is nitrogen dioxide, one of the compounds that give photochemical smog its characteristic yellowish brown color.

Only about 10% of the nitrogen compounds in the air are the result of human activity. The rest are formed by natural processes, such as the decay of organic matter. However, since the human-made 10% is emitted mostly in industrial urban areas, concentrations there can become high enough to cause concern.

HEALTH AND ENVIRONMENTAL EFFECTS
Various nitrogen oxide compounds have been implicated in human health problems. People who have chronic respiratory trouble, as well as those with kidney or heart disease, may get worse when nitrogen oxide levels are high. Nitrogen emissions can also be converted to weak nitric acid in the atmosphere, so the concern about acidic precipitation applies to nitrogen as well as to sulfur compounds. The greatest single impact of airborne nitrogen compounds, however, is their role in the reactions that produce smog and impair visibility.

As with sulfur, recent research casts doubt on previously widely held beliefs about nitrogen emissions. Investigators are still seeking answers to a number of puzzles. How do nitrogen compounds combine with other compounds in the atmosphere? Which combinations are potentially harmful? (Only a fraction of these combinations have been tested so far.) What proportion of nitrogen oxides in the air comes from power plants? From other industries? From automobiles? The answers to these questions will determine the kinds of control technology that are most appropriate. For example, if the most harmful nitrogen oxide emissions are found to come from automobiles, quite different steps will be called for than if they come from power plants.

CONTROL TECHNOLOGY   There are two major approaches to the control of nitrogen oxide emissions from coal-fired power plants: modifying the actual combustion process to achieve cleaner burning and postcombustion cleanup.

The most efficient ways of burning coal—those that produce the most energy—also produce the highest levels of nitrogen oxides. By holding down peak temperatures when the coal is burned, utilities can reduce the level of nitrogen oxide emissions. But there is always some loss of efficiency when any low-temperature technique is used.

The alternative is to clean nitrogen oxides out of the stack gases after the coal is burned. These postcombustion techniques for controlling nitrogen oxides include injection of chemical additives, such as small quantities of ammonia, into the gases. In the presence of a catalyst the resulting chemical reaction forms nitrogen gas and water. However, these postcombustion techniques tend to be expensive and to require tight temperature controls.

Particulates

Particulates are tiny particles of solid or liquid airborne material, ranging in diameter from about 0.005 micrometer for *aerosols* (particles that remain suspended in the air) to about 500 micrometers—500 millionths of a meter—for flakes of soot. About 80% of all suspended particulates are natural, in the form of dust, smoke from forest fires, sea salt, and volcanic debris. Other particulates are caused by such human activities as agricultural burning, mining, and waste incineration—and by burning fuels, especially coal. Because they are so small and light, particulates can remain suspended in the air for long periods of time and can travel immense distances—up to 24,000 miles.

HEALTH AND ENVIRONMENTAL EFFECTS

Particulates, whether they are inert or chemically active, may irritate the lungs. So they are being investigated in connection with a number of respiratory ailments, including bronchitis, asthma, emphysema, and lung cancer.

The body has its own defenses against particulates, which are usually trapped and expelled by the nose and the throat. The smaller the particulate, however, the more likely it is to slip past these defenses and find its way deep into the lungs. Up to 30% of the tiniest particles may lodge in the lungs, where they can trigger tissue changes and possibly disease.

Particulates also affect the environment. In the past, large flakes of black soot from coal burning covered the nation's industrial cities, but modern control techniques have helped erase most of this urban grime. Now the esthetic problem is mainly one of reduced visibility. High particulate concentrations can obscure the view in cities, as well as in rural areas where heavy agricultural burning or mining takes place.

There is some suspicion, too, that high particulate concentrations could actually alter our climate. They could act as a screen or filter, blocking out solar radiation. The effect would be a cooling of the earth's surface, an effect with potentially profound implications for human activities and population patterns.

Current research suggests that very fine particles—some so small they can be seen only through a microscope—are the main producers of unwanted health and environmental effects. Besides being the ones most likely to invade the lungs, they are the principal carriers of toxic trace metals in coal, such as lead, cadmium, and arsenic. And they are the ones most effective in scattering or absorbing sunlight, thereby reducing visibility and/or blocking out the sun's rays. So fine particles are the latest focus of a growing control effort.

CONTROL TECHNOLOGY   The airborne particles produced specifically from burning fuel, as opposed to the many other particle sources, are called *fly ash*. Utilities control fly ash by three basic methods: electrostatic precipitators, scrubbers, and fabric filters, or baghouses.

When the gases containing fly ash from coal combustion are passed through a device called an *electrostatic precipitator*, they are given an electric charge so that the particles adhere to collection plates. The particles are then knocked loose into collection bins. Electrostatic precipitators can be more than 99% efficient in terms of the total amount of fly ash removed, but they are much less efficient in dealing with the finer particulates. In addition, because the size of a precipitator is determined by the efficiency of fly ash collection required, an increase in efficiency requires a corresponding increase in equipment size and cost.

Scrubbers employ a liquid—usually water—to wash particles out of the flue gas. (Using scrubbers to extract sulfur oxide requires the addition of lime or other chemically reactive compounds.) Although the overall efficiency of scrubbers in particulate removal is more than 99% for larger particles, it is only about 20% effective for the smaller particles that are suspected of being the most harmful.

A *baghouse* is an installation containing thousands of heat-resistant fiberglass bags approximately 1 foot in diameter and 30 feet long. Plant emissions are passed through these bags, which in tests have trapped up to 99.9% of all particulates, including the smaller ones that evade other means of control. Whether baghouses are economically feasible is the prime question in baghouse particulate control.

Carbon Dioxide

Carbon dioxide is released into the atmosphere by natural processes, including our own respiration. However, fossil fuel combustion releases nearly 20 billion tons of additional carbon dioxide every year, and the total carbon dioxide in the atmosphere has risen by about 13% during the past 100 years. The reason for concern is carbon dioxide's possible effect on the earth's climate.

ENVIRONMENTAL EFFECTS   The atmosphere gains very little of its heat directly from sunlight. Instead, solar radiation is absorbed by the earth, converted to heat, and reradiated from the earth's surface. If this heat is then absorbed by a blanket of water vapor and gases (especially carbon dioxide) in the atmosphere, that blanket can trap the heat and hold it close to earth. The result is a net heating of the earth and its atmosphere, called the *greenhouse effect*.

The global heat balance responds to other human activities as well. Agriculture and logging, irrigation, cities, and highways all change the amount of heat the earth absorbs and reflects. In addition, the use of both fossil and nuclear fuels creates a certain amount of heat that in one way or another is ultimately dissipated into the atmosphere, where it could be trapped by a carbon dioxide blanket.

HARMFUL OR BENEFICIAL?   What would be the impact of an artificially induced warming trend? Some experts believe that the average temperature might be raised enough to melt some polar ice, flooding coastal cities and disturbing wind and rain patterns all over the globe. Other scientists believe that a warmer earth with different precipitation patterns might actually be more beneficial to agriculture and therefore to human life.

Perhaps this warming trend would even be canceled out by the cooling impact of a particulate sunscreen. In any case, many experts feel that human activity is now a significant factor in the earth's climate, and that any climate changes we create will last for a very long time.

SOLID WASTE MANAGEMENT

The more efficient the scrubbers and other control devices become in cleaning up power plant emissions, the greater the amount of solid waste that must be disposed of. Each year the electric power industry generates about 60 million tons of solid waste.

Scrubber Sludge

Most of the waste from scrubbers is a thick sludge about the consistency of toothpaste. Scrubber sludge is generally stored in huge ponds, where its 40% water content evaporates. The sheer quantity of this sludge makes its management and storage a substantial task.

In addition, these wastes may contain traces of potentially toxic materials such as arsenic, mercury, and selenium. The concern is that these materials could *leach*, or dissolve, into the surrounding soil and groundwater. If the Environmental Protection Agency (EPA) concludes that scrubber sludge is a hazardous waste, new technology will have to be developed to prevent the possibility of leaching (Figure 8-2). Such technology, which would consist of pond liners and leachate monitoring systems, might drive the costs of disposal from about $2 per ton to as high as $90 per ton. Indeed, if all utility wastes were declared hazardous, disposal costs might nearly equal fuel costs.

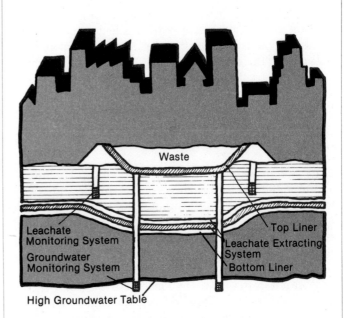

Figure 8-2. Solid Waste Management System
Under proposed federal regulations, wastes deemed hazardous to human health or the environment will be disposed of in carefully constructed landfill or pond systems. This pond model would use two liners to prevent wastes from leaching, or dissolving, into the surrounding groundwater.

Ash

Much of the solid waste produced by coal-fired power plants is in the form of mountains of ash. A plant with 100-megawatt capacity will produce about 25,000 tons of ash a year. Power plants produce enough ash each year to fill a railroad train reaching from the East Coast to the West Coast and back again—about 6000 miles.

About 13% of that ash is used in road building or is mixed with cement for other building uses. The rest is disposed of in pits or used for landfill. Research is now exploring several new options for ash disposal and recycling.

The biggest concern in solid waste management is the fear that cleaning up the air by means of accepted control technology may inadvertently threaten our water supply. Toxic wastes cleaned out of the air and stored in solid form could conceivably find their way into surface or groundwater systems. The challenge, then, is to solve today's air quality problem without creating a water quality problem for tomorrow.

WATER QUALITY

Generating electricity requires huge amounts of water to draw off and dispose of waste heat. Electric power production accounts for approximately one-fourth of all water drawn from the country's waterways. But most of this water, once used for cooling, is returned to the waterways from which it was drawn. Actual consumption of water by utilities, due mostly to losses by evaporation, comes to only about 1% of the national total.

Cooling Systems

Many systems exist for power plant cooling. The most common, the once-through system, is simple and economical but requires large amounts of water. Other methods, such as cooling ponds and evaporative cooling towers, have different advantages and drawbacks, but all require the use of water, a resource that is already in short supply in certain regions of the country.

Research is under way to conserve fresh water by using waste water reclaimed from municipal, agricultural, or industrial users for power plant cooling. Other research endeavors to develop an improved dry-cooling system based on the condensation and evaporation of ammonia. However, such a system is not expected to be commercially feasible in places where water is available, since it would raise the cost of energy generation by 10% or more over that of plants that use conventional wet-cooling systems.

Chemical Contamination

As we have seen, the leaching of toxic materials from storage ponds and pits into groundwater is one form of water pollution that may result from electricity production. Another potential problem is that various hazardous substances may dissolve in the working water while it is in the plant and remain in the plant's water discharge streams. For example, arsenic condensed on fly ash may dissolve in sluice water. And chemical additives such as chromium, commonly used for corrosion control, may find their way into the plant's liquid emissions as well. Work is now under way to devise a reliable and inexpensive method of reducing the amount of these trace elements in power plant discharge streams.

Chlorine discharge is also the subject of research. Chlorine is added to the cooling water to prevent the growth of biological organisms in the system. But there is concern that the presence of chlorine and its com-

Table 8-1. Potential Uses for Low-Grade Heat
The heat that power plants throw off as a by-product of electricity generation can be useful for a variety of other purposes. Making use of this reject heat is one way to conserve energy resources by using them more efficiently.

Area	Use
Agriculture	Frost control
	Greenhouse heating
	Soil warming to extend growing season
Aquaculture	Growth-rate enhancement
	Spawning control
Industry	Chemical processing
	Integrated energy systems
	Sewage treatment
Recreation	Bathing beaches
	Warm water fisheries
Space heating	Development of new cities
	Retrofitting existing developments

pounds could upset the ecological balance outside the power plant if it is not removed or greatly reduced before the water is discharged.

The Heat Problem

Yet another problem is heat, or thermal pollution. Federal regulations control the discharge of warm water because the heat is suspected of altering streams in ways that can kill fish or weaken them so that they are vulnerable to parasites and disease. Disposing of the heat that accompanies power generation without threatening the environment is a concern with virtually every sort of power option, including the developing technologies such as solar electricity.

Low-grade waste heat can have many beneficial uses, however, particularly in agriculture and aquaculture. For example, a full-scale project is now demonstrating how hot waste water from a power plant can be piped directly to nearby greenhouses for soil heating. It is estimated that the waste heat from a 1000-megawatt power plant could heat several hundred acres of greenhouses. Table 8-1 gives some other potential uses for waste heat.

HIGH VOLTAGE

Overhead transmission lines are the superhighways of the electric utility industry. As the demand for electric power increases, so does transmission voltage. Higher-voltage lines can carry more electricity with greater efficiency, with smaller expenditures of money and land.

Concern has risen recently over the possible biological effects of exposure to the electric fields of high-voltage transmission lines. Reports from the Soviet Union describe such symptoms as loss of appetite and energy and diminished sex drive among switchyard workers who were exposed for long periods to high-voltage electric fields. Similar reports have come from Spain. However, studies of workers under similar conditions in the United States, France, West Germany, and Sweden reveal no physical or biochemical effects. Intensive ongoing studies of workers, as well as experiments with large and small animals, are being conducted to isolate any possible response to prolonged electric field exposure.

RADIATION

Because radiation has been examined so intensively and because its effects are easier to isolate than those of, say, airborne sulfur compounds, more is known about the health effects of radioactive emissions than about most other substances now under study.

Radiation is energy that travels in rays or particles. We are born into a world of radiation, the most familiar example being ordinary light. There are two types of radiation: *ionizing* and *nonionizing*. Although over-exposure to nonionizing radiation, such as light and sound, can cause discomfort and even impair sight and hearing, it is not considered life-threatening. Ionizing radiation, on the other hand, involves electrically charged particles so energetic that they can disrupt vital body functions. The natural background radiation of our planet is ionizing radiation; so are X rays and the radiation produced in a nuclear power plant.

Natural and Human-Produced Sources

Naturally occurring radiation from the water, the soil, the air, cosmic rays, and even our own bodies exposes every person on earth to an average of just over 100 millirems every year (Figure 8-3). (A *rem* is a unit of radiation; a *millirem* is one-thousandth of a rem.) The amount of natural radiation to which any individual is exposed depends on such factors as the type of rock in the surrounding area and the kind of house he or she

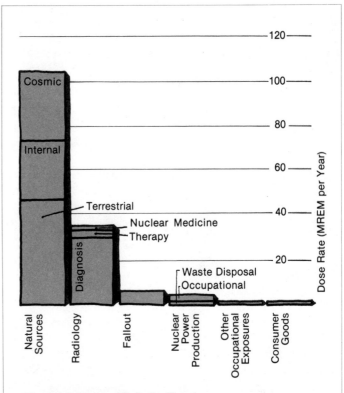

Figure 8-3. Sources of Radiation Exposure
Nature is the main source of radiation exposure. Of human-made sources, medical X rays top the list, with weapons fallout and on-the-job exposures playing lesser roles. The average citizen's exposure to radiation from nuclear power production is quite small compared with exposure from these other sources.

lives in. Even altitude is important, since the atmosphere screens out a certain percentage of cosmic rays. People living at sea level receive less exposure than those living at higher altitudes (Table 8-2).

Sources of radiation produced by human activity add an average of another 100 millirems per year. By far the most comes from medical X rays, with diagnostic and therapeutic use of X rays accounting for more than 90% of our radiation exposure from these sources. Weapons fallout and on-the-job exposures also play a role. The contribution of nuclear power plants is comparatively small.

Table 8-2. Calculation of Annual Radiation Exposure
Annual radiation exposure can vary from person to person,
depending on where and how one lives. Living in the mountains,
taking jet flights, and watching a lot of television can all increase
the dosage, although not to levels that are generally considered
dangerous. The average person in the United States receives
200 millirems a year.

	Common Source of Radiation	Your Annual Inventory (mrem per year)
Where you live	Location: cosmic radiation at sea level	40
	Add 1 for every 100 feet of elevation where you live	_____
	House construction: wood 35	
	concrete 50	
	brick 75	
	stone 70	_____
	Ground (U.S. average)	56
What you eat, drink, and breathe	Water and food (U.S. average)	25
	Air (U.S. average)	5
How you live	Jet airplanes: number of 6000-mile flights × 4	_____
	Radium dial wristwatch: add 2	_____
	Television viewing:	
	black and white: number of hours per day × 1	_____
	color: number of hours per day × 2	_____
	X-ray diagnosis and treatment	
	limb X ray: 420	
	chest X ray: 150	
	stomach X ray: 350	
	colon X ray: 450	
	head X ray: 50	
	spinal X ray: 250	
	gastrointestinal tract X ray: 2000	
	dental X ray: 20	_____
How close you live to a nuclear plant	At site boundary: number of hours per day × 0.2	_____
	One mile away: number of hours per day × 0.02	_____
	Five miles away: number of hours per day × 0.002	_____
	TOTAL	_____

The average person is exposed to about 200 millirems of ionizing radiation a year. How much exposure is harmful? A massive dose above 600,000 millirems would probably prove fatal within a month. Below 25,000 millirems, on the other hand, there are no immediate and observable effects. Delayed effects can occur, the most serious being cancer or genetic damage that may show up as many as 20 or 30 years later. But despite considerable disagreement among radiation scientists as to the exact dose-response mechanism at work in low-level exposures, there is general agreement that any such effects in an exposed population will be very small.

The Nuclear Contribution

What is the contribution of nuclear power plants in the United States to the average exposure of 200 millirems per person per year?

Civilian nuclear power's contribution to average population exposure is much smaller than that of normal background radiation—probably less than 1 millirem per year. Government standards limit annual exposure for people who live near a power plant to 10 millirems per year. The Nuclear Regulatory Commission standard of safety for workers in nuclear plants is 5 rems a year and not more than 3 rems in 3 months.

Even with those strict standards, radioactive discharges from nuclear power plants are generally from 10 to 100 times below—safer than—permissible levels. Under normal operating conditions, these limits are never even approached, and the contribution of a nuclear power plant to radiation levels is often difficult to measure because it is overwhelmed by natural background radiation. In fact, because of the radium and uranium found in coal, coal-fired power plants emit about the same amount of radioactivity as nuclear plants.

Under normal conditions, then, the operation of a nuclear power plant does not add significantly to the average amount of radiation received by the American citizen, even one who works in such a plant.

What can be said about the possibility of accidental high-level radiation releases from a commercial nuclear plant? In fact, no such releases have occurred, and some experts believe that we do not have enough experience yet to be able to make accurate long-term predictions as to the likelihood of such an event.

The March 1979 reactor malfunction at Three Mile Island was the most serious accident involving a civilian nuclear facility in the United States, and the actual exposure there averaged only about 1.5 millirems for people within a 50-mile radius. Of course, those closest to the plant received higher-than-average exposures, sometimes as much as 80 millirems. This level of exposure is comparable to that from certain types of medical X rays.

The United States has experienced no high-level releases of radiation and no reactor meltdown in more than 450 reactor-years of peaceful nuclear power use. (A *reactor year* means one nuclear reactor in operation for one year. Ten reactor years might mean one reactor in operation for ten years, or ten reactors operating for one year.) This safety record reflects the elaborate precautions engineered into nuclear power plants as part of their basic design. These precautions are expected to grow even more effective as intensive new research in response to the events at Three Mile Island focuses on ways to improve nuclear safety, from improved monitoring systems to better operator training.

Waste Disposal

Nuclear wastes also emit radiation. Whether these wastes can be handled without serious risk to human health and the environment is a topic of widespread concern. Several states have banned new nuclear plants until the federal government demonstrates that it is possible to dispose of the wastes safely.

The principal waste from a 1000-megawatt coal-fired power plant is carbon dioxide, which is emitted at a rate of about 10 pounds per second, and ash residue, produced at about 30 pounds per second. In contrast, the typical large nuclear reactor produces only 30 to 40 tons of waste per year. And once it has been prepared for disposal, the annual volume of waste from a 1000-megawatt nuclear reactor would be about 2 cubic meters.

The technology for processing waste has already been developed in several countries and is now entering large-scale production in France. First, the spent nuclear fuel is stored in liquid for several months to allow some of the radioactive products to decay. Then, through solvent extraction, the fuel is reprocessed. The uranium and plutonium can be separated out at this point for reuse as fuel. Everything remaining is high-level radioactive waste, which is then heated to drive off the liquid. A fine

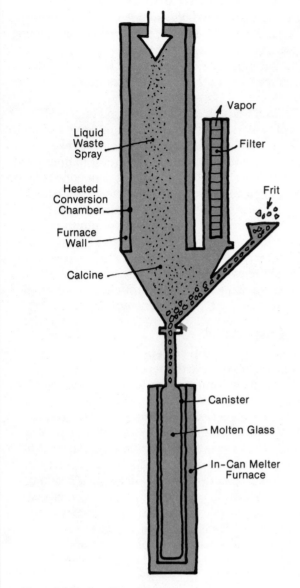

Figure 8-4. Nuclear Waste Disposal
Liquid radioactive waste is first heated in a conversion chamber until only a fine powder called calcine is left. Then frit—material for making glass—is added to the calcine, and the two are fused into a solid block of glass right inside the stainless steel canister that will be used for burial.

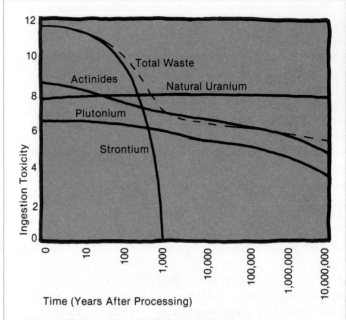

Figure 8-5. Decay Times for Radioactive Wastes
The hazardous parts of nuclear waste eventually decay to a level of radioactivity lower than that of natural uranium lying in the earth. For example, the strontium in nuclear waste becomes less toxic to humans than natural uranium ore in about 450 years. The total waste, including plutonium, takes from 500 to 1000 years to reach that same level.

powder called *calcine* is left. It takes up only about one-eighth as much space as the liquid waste did. The calcine is fused into molten glass directly in the stainless steel canister to be used for disposal (Figure 8-4).

The plan is to store these canisters in vaults cut into stable geologic formations more than 1000 meters (3280 feet) deep. Because the storage vault would not be near any groundwater; because glass is highly resistant to leaching by water, as well as to other forms of decay; and because the surrounding rock would absorb any leaching that might occur, it is estimated that the long-term hazard from radioactive waste disposed of in this manner is about the same as that from naturally occurring uranium and radium in the earth (Figure 8-5).

At present, spent fuel from commercial reactors is being stored in water-filled storage pools at each nuclear plant. The government is considering construction of a $300 million water-filled storage facility for spent fuels that could ease the problem for 50 to 100 years. More important, plans for permanent underground disposal are proceeding, using comparative surveys to determine which of several possible sites in the nation would be most suitable.

Actually, radioactive waste from commercial power plants is a very small part of the nuclear waste problem. Most wastes come from defense uses: bombs and nuclear-powered submarines. At this point, the amount of highly radioactive accumulated defense waste awaiting permanent disposal in tanks and burial pits is about 500,000 tons. Commercial nuclear reactors, in contrast, have produced less than 5000 tons of spent fuel, which can be processed into a much smaller volume. So military wastes are about 100 times greater than those from civilian nuclear power use.

It is clear that a plan will have to be carried out, and soon, for dealing with the problem of radioactive waste disposal. But the contribution of nuclear power plants to that problem is relatively minor.

EFFECTS OF OTHER TECHNOLOGIES

Other electric power generation technologies affect the environment. Oil combustion releases carbon monoxide, carbon dioxide, nitrogen oxides, sulfur dioxide, and particulates. Of all the fossil fuels, natural gas is the cleanest-burning and presents the fewest pollution problems. Yet no known source of electric power generation is completely problem-free. For example, if solar satellites were used to collect the sun's energy, they would beam that energy to the earth in the form of microwaves. The effects of microwave transmission are not fully known, but they are suspected of having negative effects on human and animal health.

In hydroelectric power generation, the damming of rivers has a long-term ecological impact. The buildup of silt and sediment has proved to be a stubborn environmental problem at many sites. Dam safety is also a serious concern because of the potential impact of dam breaks and consequent flooding.

Problems with geothermal generation include thermal pollution, mineral pollution (waste water from geothermal energy generation has high mineral content), and the release of noxious natural gases, such as sulfur, ammonia, and hydrogen sulfate, into the atmosphere.

POLLUTION CONTROL COSTS

There are no entirely nonpolluting, risk-free methods for generating the electricity we depend on. What we must determine, then, is how much we have to spend on the control technology to minimize the risks. In effect, we must decide on how much insurance we should buy.

Dollar Costs

The electric utility industry spends nearly $2.5 billion a year on new plants and equipment for the control of air, water, and solid waste pollution. Compliance with new Environmental Protection Agency rulings on air pollution control may well cost another $200 billion by the year 2000. The cost of scrubbers alone is expected to add between 18% and 35% to the cost of building and operating a coal-fired power plant.

Energy Costs

The energy penalty of pollution control is also high. For example, up to 3% of a coal-fired power plant's total output is required for the operation of stack gas scrubbers alone. Approximately 10% of all electric energy produced in the United States is used to run environmental cleanup equipment. And the stricter the rules and the more sophisticated the equipment needed, the higher the energy costs will be.

That is the energy actually consumed by environmental cleanup requirements. In addition, there is the energy that is never even produced because the accompanying dollar costs for cleanup would be too high. So the energy penalty works both ways: some energy is actually consumed to run pollution control equipment, and some potential energy is never developed because cleanup costs make it uneconomic to do so.

Amount of Cleanup

The dollar cost of removing 80–90% of the sulfur from coal emissions generally ranges from $0.50 to $1.00 per pound. To remove more than 90%, however, the cost skyrockets to about $5.00 per pound. This problem is common with pollution control: capturing the last fraction of the substance to be controlled is often far more difficult and expensive than capturing the first 95%. When is it necessary to go for the last fraction? When is it worthwhile? These are questions that industry and government alike are seeking to answer.

Whenever a utility must install new equipment to meet more stringent emissions standards, that cost will ultimately be passed along to the consumer. The problem is deciding where to draw the line between necessary insurance against real hazards and unnecessary insurance against remote risks. There is no doubt that uncontrolled emissions from power plants could pose a real hazard to human health and the environment. That is why such emissions are already tightly controlled. The question that remains is who should decide whether we should spend more money tightening restrictions on substances that may already be adequately controlled or even overcontrolled.

ENVIRONMENTAL LEGISLATION

If the technological and economic problems of pollution control seem complicated, the legal side does little to simplify the situation. Emissions *standards* prescribe the legal limits on pollutants that a given source, such as a car or a power plant, can release. Emissions in excess of the standard violate the law. Standards are supposed to reflect a "safe" threshold. The difficulty is that although most plant emissions are indeed believed to have such a threshold, researchers still do not know enough about these substances to know exactly where that threshold lies. So standards can change as new knowledge emerges.

Air Quality Legislation

Air pollution, for example, was treated as a local problem for many years. Legislation and control efforts took place at the city, county, or state level and concentrated mostly on visible pollutants, such as black smoke. With the use of electrostatic precipitators and other control devices, the black smoke that had begrimed America's industrial cities during the early 1900s was virtually erased by 1950.

As research during the 1960s revealed how far air pollutants could actually travel, the problem was gradually recognized as a national one. In 1963 the Clean Air Act gave limited enforcement powers to the federal government, although the states were still responsible for setting their own standards and for requesting federal enforcement assistance. In 1967 the Air Quality Act required that the Department of Health, Education, and Welfare (HEW) issue criteria and recommend control techniques for six major pollutants. However, each state was still responsible for setting its own air quality standards, based on HEW criteria.

In 1970 the Clean Air Act Amendments finally mandated uniform national air quality standards. Rather than starting from the standpoint of what was technically feasible, the government made a decision as to what standards it believed necessary to protect the public health and the environment and then required industry to meet those standards. In other words, industry would have to create new technology to comply with tougher government standards. This approach is known as *control technology forcing*.

The latest federal air quality laws, the Clean Air Act Amendments of 1977, have once again lowered the allowable atmospheric levels of sulfur dioxide, nitrogen dioxide, and particulates. Many experts now believe that these standards exceed the limits necessary to protect human health and the environment.

The process of setting air quality standards is a series of continuing adjustments, as research sheds new light on the behavior and effects of various airborne substances. Revisions of the law can be slow in coming, however. And the question always remains: How close are we coming to an ideal balance between effectiveness and cost in pollution control?

Decision-Making Conflicts

In 1970 the EPA was created to set and enforce pollution standards. But there are still a number of federal agencies with overlapping responsibility for shaping environmental policy. State and local agencies are involved, too. Various industries have a voice, and so do the many special interest groups that lobby for and against particular pieces of environmental legislation. The result is that many groups have a hand in shaping environmental policy.

There are also many priorities. Some of the leading ones are health and environmental protection, inflation control, and national energy needs. These priorities are not always in harmony with one another. For example, although a certain amount of environmental damage is inevitable in strip-mining coal, the coal that is mined can help meet national energy needs and reduce our dependence on foreign oil. If we try to honor both priorities by restoring the land after strip-mining, then inflation control is threatened because restoration can be very expensive and the costs will be passed along to consumers in the price of the coal. How do we reconcile all these objectives?

Chapter 9

MAKING ENERGY DECISIONS

Energy use is related to every aspect of our society: economic growth, population distribution, health, transportation, productivity, communication systems, land use, air and water quality, employment, and foreign policy. They are all key variables in a complex system, and they all affect each other directly or indirectly. In such complex relationships, establishing goals and making decisions cannot be simple.

Most of us would agree, however, that among our goals would be to have energy that is as inexpensive as possible, reliable, and environmentally acceptable. These goals apply to energy use in general and to electricity use in particular. They help shape energy decisions made on many levels—by government policymakers, by utility and other industry planners, by individual consumers. We all make decisions that will affect the nation's energy future.

In making our energy decisions, we must keep in mind two important questions: How much energy will we need? What balance must we strike in order to get it? Every institution, group, and individual will view these questions in a slightly different way. What we all have in common is the need to look ahead and consider the long-range consequences of today's decisions.

THE UNCERTAINTY FACTOR

The first step in planning for any energy supply is to estimate the amount of energy that is going to be needed. In the past, predicting energy demand was a relatively easy task. Demand grew steadily, methods for providing energy were reliable, and environmental effects had not yet become a social issue. Forecasting appeared to be a simple matter of straight-line projections. However, there have been dramatic changes in the factors that affect energy use.

Foreign Supplies

International politics have changed traditional energy supply and demand. World oil prices have soared, and no one knows how prices will fluctuate in the future. Because fuel prices affect the demand for electricity and other energy forms, price uncertainty creates demand uncertainty.

Conservation

Higher prices increase conservation efforts, but no one knows how much or for how long. How much less would people drive their cars if gasoline cost $2.50 a gallon? Would they drive less at first and then gradually venture onto the highways more and more as they became accustomed to higher prices? How would conservation affect economic growth? Can we decrease our use of expensive fuels and maintain total energy use?

Utility Costs

Another uncertainty is utilities' ability to meet rising costs. In addition to skyrocketing fuel prices, the costs of plant construction—both the high interest rates on financing and the actual costs of land, labor, and materials—are escalating. So are the costs of operating and maintaining a plant once it is built.

The higher these costs go, the higher utility rates must go to cover them. As these costs are passed on to customers, their likely response will be to cut back consumption. How great will that cutback be? How much will rising costs—and the resulting higher rates—trim the demand for electricity or increase the efficiency of its use?

New Technologies

Technology has brought us convenience, health and safety, and wealth. Many people are confident that technology will provide a solution to our energy problems as well. Unlike the limited mineral resources that nature provides, technology is a human resource whose abundance and variety can be increased continuously. Yet this very potential for growth and change adds to our uncertainty in making projections, for the form and extent of new energy-producing systems are difficult to foresee. How fast will technologies develop? When will the breakthroughs come? How will new technologies affect existing ones?

Public Policy

Public support for future energy development is also uncertain. Utility, industrial, regulatory, and special interest groups often have conflicting views, which can result in delay of an effective energy policy.

Energy legislation generally has two goals. The first is to increase the production rate from domestic fuel resources—for example, by removing price controls on producers, by providing government funding for research, or by shortening the licensing time for power plants. The second goal is to decrease energy use—through mileage standards imposed on automakers, through tax incentives for insulation, through efficiency standards for appliances, and through other conservation measures. Many such policies are newly established, and others are still being formulated. Because we do not yet know what legislation will be in effect over the next 20 years, it is difficult to estimate how well energy policy will succeed in speeding production or slowing demand.

Forecasts of future energy demand, then, must be subject to constant revision. For utilities, these changing forecasts will affect capacity expansion decisions. Utility systems go through several stages—studies, design, licensing, construction, and startup—and at each stage some decision is made regarding the feasibility and schedule for proceeding to the next. These decisions depend on factors that are continually changing, so uncertainty affects planning all along the way.

COPING WITH UNCERTAINTY

Planning for higher demand than actually occurs will likely result in an excess of production capacity. Planning for a lower demand than actually occurs will probably result in an energy shortfall—the failure to supply an adequate amount of energy when it is needed. For both producers and consumers of electric energy, various costs are associated with these two extremes.

Some costs go up when there is too much electricity-producing capacity, and others go up when there is too little. For example, too much capacity is costly because power plants are expensive to build and maintain. On the other hand, new plants are often more economical to operate than older ones (such as oil-fired plants), so added capacity can make for a more economical generating mix.

Environmental costs can either shrink or grow with an increase in capacity. An undersized, older utility system has plants that are being pushed past their highest efficiency. As a result, environmental control devices may not work as well as they should. A system with planned overcapacity can schedule proper maintenance to keep it environmentally sound. An oversized system does not necessarily produce more pollution than an undersized one. If the level of power production is the same, the larger plant may actually produce a smaller total amount of pollution. The environmental drawback to overcapacity is simply that it uses land that could be used for other purposes.

Outage Costs

A major cost, and one often overlooked, is the cost of power outages when capacity is insufficient to meet demand. We all know that blackouts and brownouts are annoying and even dangerous, but we tend to forget how expensive they can be. When the meat packer's refrigeration units fail or the manufacturer's assembly lines stop, the consumer eventually pays. The costs of a prolonged power outage are eventually passed on to the public when businesses try to recover their losses by raising prices. Outage costs can also run very high in lost jobs, which must be considered when a utility tries to plan a least-cost system for its customers.

Striking a Balance

The task that utility planners face is to find a balance between surplus and deficit—that point at which the lowest cost occurs. And that means determining just the

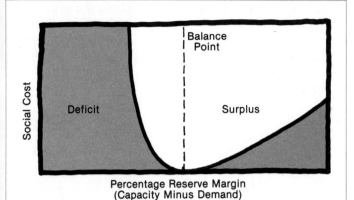

Figure 9–1. Social Costs of a Supply-Demand Mismatch
The disruptions triggered by past energy shortages suggest that it is cheaper for the nation to have too much power-producing capacity than too little. Building and maintaining idle power plants is expensive, but blackouts and job losses related to insufficient power can be even more expensive. If capacity falls behind demand, it can take 10 years to catch up.

right amount of insurance to buy in the form of generating reserves. The problem, of course, is that planners can never be sure exactly how much power the public will demand. Many energy planners think that it is cheaper to decide on a higher reserve margin than on a low one (Figure 9–1).

If a utility maintains a low planning reserve margin and demand turns out to be consistently higher, over time, than was anticipated, the utility must attempt to meet that demand by installing modes of electricity generation with shorter lead times (which tend to use expensive fuel), by interrupting service, or even by incurring outages. The resulting costs to consumers are greater than the costs of supporting a higher planning reserve margin, even though that higher planning margin entails the commitment to begin the studies, licensing, and construction of generating units that won't actually be needed if demand slows down.

The costs, like the curve in Figure 9–1, are asymmetrical. For example, it would cost the consumer more if the local utility were to build 20% too little capacity than if it were to build 20% too much.

Flexibility

Flexibility is important. Uncertainty cannot be controlled, but that does not mean that it can be ignored.

Utilities have to plan around uncertainty in a way that allows a swift response to changes in demand and supply. The more easily and quickly a utility can adjust its capacity expansion plans in accordance with new information, the lower its costs—and consumers' costs—will be. Because predicting demand must always be a matter of guesswork, the point of staying flexible is to minimize the costs of being wrong.

Consequences of Undercapacity

If surging demand for electricity starts to exceed production capacity and there are no backup utility plants ready to swing into action, there is no way to create an instant increase in capacity. It takes time for plants to be designed, licensed, constructed, and brought on-line. The amount of time varies according to the type of plant and the particular utility. On the basis of average lead times for all plant types, it is estimated that it would take about 10 years to correct an underplanning error. Meanwhile, society itself would be affected to a greater or lesser extent by the scarcity of power.

Over a long period, these effects could be severe. For example, if there were insufficient energy to run industrial plants, there would be ample time for the production of goods to drop and for the prices of these now-scarce goods to soar. Layoffs could last for years. And there would be plenty of time for the U.S. energy shortfall to shake the world economy, reverberating throughout the structure of international trade and politics.

Consequences of Overcapacity

Overcapacity, on the other hand, is a much more easily reversible error. Schedules for the construction of power plants can be interrupted and changed (Figure 9–2). If the design phase of a particular plant has been completed and the demand forecast suddenly goes down, the plant need not be built at all. When and if demand rises, it will not be necessary to spend time on design before beginning plant construction. Even after both design and licensing are completed, the plans for a plant can be shelved.

Thus, power producers can adapt relatively easily to a change in demand, as long as early stages of new capacity are already complete. Overexpansion can be

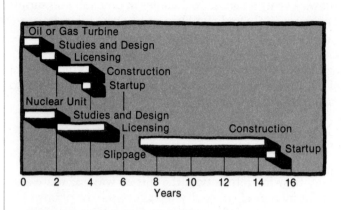

Figure 9–2. Timing of Capacity Additions
Each type of generating unit—a turbine running on oil or gas,
for example, in contrast with a baseload nuclear unit—has its own
typical schedule. If predicted demand decreases, work on the unit
can be stopped. If demand increases, work can resume, or the
unit can be replaced by a type that has a shorter lead time.

Here are some of the questions that we can ask ourselves to help make sensible decisions about the nation's
energy future:

- What are the nation's growth prospects—in population, GNP, jobs?
- How much energy will we need to support a particular level of growth?
- What forms of energy—oil, gas, electricity—will be
  most in demand?
- What level of fuel importation does this demand profile imply?
- How can we best conserve energy and use it more
  efficiently? What role will higher prices, technological
  advances, government regulation, and new personal
  values have in promoting conservation?
- How much can we boost domestic energy production
  from any or all of our existing sources, such as oil,
  gas, coal, nuclear, and renewables?
- What contribution can we expect from emerging
  energy sources, such as solar, wind, or biomass? How
  soon will that contribution be made?
- When will we be able to store electricity in much
  the same way we store oil and gas? How could this
  ability affect our prospects for power from the sun,
  the wind, and other intermittent sources?
- How does energy production of various types affect
  the environment?
- How much are we willing to pay—in dollars and
  in energy either consumed or never produced—to
  achieve a certain level of environmental protection?
- How much energy capacity do we need?
- How much do we need to hold in reserve?

Each of these questions invites—demands—further
study. All current and emerging energy options must be
considered, for we are going to need every viable energy
source we can find. The decision-making process will
have to be structured in such a way as to pool our
knowledge and resources, minimize conflict, and maintain enough flexibility that we can alter our course as
new information becomes available. If we want to live
comfortably, productively, and in harmony with our
environment, we must make conscious decisions about
our energy future.

slowed down, but underexpansion takes so long to speed
up that it may never catch up with need.

Timing is critical in energy planning. Because of the
long lead times necessary for capacity expansion, decisions must be made well in advance of actual need. The
decisions that are made today will determine whether
we as a nation have too much, too little, or just about
the right amount of energy at the turn of the century.

How can we make decisions when we do not have all
the facts? The answer is that all the facts relevant to
energy decisions can *never* be known in advance. We
can never be sure that all the reasonable options have
been explored or that the various options will have the
costs and benefits anticipated, but decisions have to be
made nevertheless. Failure to make a decision is itself
a decision—with consequences.

GLOSSARY

A

acidic precipitation (8) dilute solutions of sulfuric acid that precipitate as rain; possibly formed from sulfur in the atmosphere

aerosol (8) a suspension of particles in the air; a type of particulate about 0.005 micrometer in size

alternating current (7) electric current that reverses direction periodically, usually several times a second; the type of current generally produced and consumed in the United States

anaerobic fermentation (6) a biomass conversion process in which vegetation is converted into synthetic fuel by bacteria in the absence of oxygen

ash (8) a solid waste produced by coal-fired power plants

availability (3) the quality or state of being accessible or obtainable

B

baghouse (8) an installation containing thousands of heat-resistant fiberglass bags to control particulate emissions

baseload (4) a utility's minimum load over a given period of time; a baseload generating plant is designed for continuous operation and generates electricity around the clock

battery (7) a device for storing small amounts of electricity by means of a chemical reaction; the chemical energy is converted into electricity

biomass (6) organic material that can provide heat

blanket (6) one of two fuel regions in a liquid metal fast breeder reactor; the blanket contains only uranium-238

boiler (5) a large vessel that contains an assembly of tubes in which water is heated to steam to drive a turbine

boiling water reactor (5) a light water reactor in which the heat from the fuel rods causes the cooling water to boil; the resulting steam turns the turbine

bottoming cycle (5) the lower-temperature cycle in a combined-cycle system

breeder (6) a nuclear reactor that produces more fissionable material than it consumes during the fission process

British thermal unit (2) a measure of energy consumption that represents the amount of heat required to raise the temperature of 1 pound of water by 1 degree Fahrenheit

C

calcine (8) a fine powder that results when radioactive waste is heated to remove any liquid

carbon dioxide (8) a gas composed of carbon and oxygen that is released into the atmosphere by natural processes, such as our own respiration, and by such other processes as fossil fuel combustion

catalyst (6) a substance that promotes a chemical reaction without changing itself

central receiver system (6) a solar-thermal conversion system that consists of a tower surrounded by tracking mirrors, which reflect the sun's light to a thermal receiver at the top of the tower

chain reaction (5) a self-sustaining nuclear reaction yielding energy or products that cause additional reactions of the same kind

chemical contamination (8) the introduction of hazardous substances into a system such as a groundwater or river system

chemical energy (1) the energy stored in a substance's chemical composition; the energy can be released by changing the substance's composition

chemical reduction (6) a biomass conversion process in which vegetation is heated in the presence of water, carbon monoxide, and chemical catalysts

coal (3) a black, solid fossil fuel formed over millions of years by animal and vegetable matter

cogeneration (4) the production of heat and electricity from a common source

combined cycle (5) an electricity generating system that combines a gas turbine and a steam turbine

compressed-air storage (7) a storage system in which off-peak power is used to compress air, which is later used to drive a gas turbine during peak demand

conductor (7) a wire through which electricity flows

conservation (2, 9) planned management of natural resources through more efficient use

control rod (5) a major component of light water reactors that contains substances to slow down the chain reaction by absorbing neutrons

control technology forcing (8) public policy that requires industry to create new technology to meet public health and environmental standards

conversion (1, 4) the changing of one energy form to another

coolant (5) a fluid (usually water) used for cooling any part of a power plant in which heat is generated

cooling system (8) the system that carries the coolant through the power plant

core (6) one of two fuel regions in a liquid metal fast breeder reactor; the core generally contains a fuel mix of about 15–20% plutonium-239 and 80–85% uranium-238; the core is surrounded by a blanket of uranium-238

corona (7) a halolike glow that results when a conductor discharges electricity into the atmosphere

cost (9) outlay or expenditure; the amount of capital, labor, materials, and energy required by utilities to generate electricity

current (1) the flow of charged particles through a conductive material

D

damper (7) a metal weight attached to high-voltage lines to prevent or lessen vibration

demand forecast (2) a projection about how much energy or, in particular, electricity will be needed in the future

direct current (7) electric current that flows in one direction

dispersed system (6) a solar-thermal conversion system that consists of individual parabolic mirrors that concentrate the sun's light on a focal line that runs through each mirror instead of on a central tower

distribution (7) the delivery of electricity from distribution substations to individual consumers

dry steam (5) a form of geothermal energy that results when water in underground reservoirs is heated by hot, porous rock to steam, which is vented at the earth's surface, often as spectacular geysers

E

economic dispatch (4) the procedure the utility dispatcher uses to call different generating units into service

economy (2) a system of production, distribution, and consumption of goods and services

efficiency (4) the ratio of output to input

electric energy (1) the flow of charged particles

electricity demand (2, 9) the amount of electricity consumed

electrolysis (7) the chemical change of water into hydrogen and oxygen by applying an electric current

electrolyte (6) an electrically conductive chemical medium

electromagnet (7) an iron cylinder or doughnut shape that is wound with wires that carry electricity

electrostatic precipitator (8) a device for particulate control that gives fly ash particles in combustion gases an electric charge so that the particles adhere to collection plates instead of being emitted into the atmosphere

emission standard (8) an ordinance that limits the amount of pollutants a given source may release

energy (1) the capacity to do work

energy demand (2, 9) the amount of energy consumed

environment (8) all conditions and influences that affect the development and life of an organism; the ecological community

environmental effects (3, 8) the effects that the products of human activity have on the environment

F

fabric filter (8) a fiberglass bag that collects particulates as power plant emissions pass through it

flashed steam (6) an emerging geothermal technology in which hot water is suddenly depressurized (flashed) to turn it to steam for use as a working fluid in a steam turbine

fluidized bed (6) a concentrated suspension of crushed limestone in a flow of hot gas, usually air, in which pulverized coal is burned

flexibility (9) the capability of responding to changing or new situations

fly ash (8) airborne particles that are produced by burning fuel

flywheel (7) a wheel that can absorb, store, and then release energy

fossil fuel (3, 5) a fuel derived from once-living organisms

fuel cell (6) a device that converts the chemical energy of fuels directly into electricity

G

galloping (7) extreme vibration in high-voltage lines

galloping detuner (7) a device to prevent galloping

gasification (6) a method of producing a clean, burnable gas from coal

generator (4) a device for converting mechanical energy into electric energy

geopressured zone (6) a highly pressurized reservoir of hot salt water and methane gas trapped in a formation of impermeable shale

geothermal energy (6) heat energy buried beneath the earth's surface

geothermal power (3, 5) electricity generated from geothermal energy resources

greenhouse effect (8) a net heating of the earth and its atmosphere resulting from the earth's absorption of the sun's radiation, which is converted to heat and reradiated from the earth's surface to the atmosphere

grid (7) the whole interlocking system of electricity delivery, from generator to customer

gross national product (2) the total value of all goods and services produced in a nation in one year

H

head (5) the vertical distance between the water level in the reservoir behind a dam and the turbine below

heat energy (1) a form of energy that causes a substance's temperature to rise and its molecules to move faster and faster

heliostat (6) a two-axis tracking mirror used to reflect the sun's light onto a thermal receiver at the top of a tower

high voltage (8) a voltage on the order of thousands of volts

hydroelectric power (3, 5) electricity generated by the kinetic energy of falling water that passes through the conduit of a dam and turns the blades of a turbine

hydrogenation (6) catalytic reaction of hydrogen with other compounds; a step in coal liquefaction in which coal is dissolved

I

incandescence (1) light produced by the steady burning of a filament heated white-hot by an electric current

inertial confinement (6) a fusion process in which tritium-deuterium pellets are struck by laser or particle beams to induce the fusion reaction

insulation (7) material having high electrical resistivity and therefore preventing possible contact between conductors

intermediate load (4) the portion of utility load that varies daily

ionization (6) a process by which electrons, which carry negative electric charges, dissociate from the normally neutral atom, leaving a positively charged particle, or ion

isotope (5) any of two or more atoms of a chemical element with the same atomic number but different atomic mass or mass number and different physical properties

K

kilowatt (2) 1000 watts; a measure of the rate of electricity generation or consumption

kilowatthour (2) the energy that will be expended by using 1000 watts of electricity for one hour

kinetic energy (1) the energy of matter in motion

L

leaching (8) the dissolving of substances through a permeable substance; for example, the dissolving of chemicals from holding ponds into the surrounding soil and groundwater

lead time (4) the years that elapse between the time utility planners decide to build a new power plant and the time the plant actually begins to produce electricity

light water reactor (5) a nuclear fission reactor in which ordinary water serves as both moderator and coolant

liquefaction (6) a process by which coal is converted into a clean-burning liquid fuel

load (4) the sum of all a utility's customers' demand

load factor (4) the ratio of a utility's average load to its peak load, expressed as a percentage

load management (4) the promotion of maximum efficiency in the use of installed generating capacity; utility efforts to encourage customer off-peak use of electricity

M

magma (5) molten rock material

magnetic confinement (6) a fusion process in which tritium and deuterium are heated to a plasma, which is confined in a chamber by means of electromagnetic forces

magnetohydrodynamics (6) a generating process in which the heat energy of a hot fluid is converted directly to electric energy

mechanical energy (1) energy expended by the application of force to an object, causing the object to move

millirem (8) a unit of dose of radiation equal to one-thousandth of a rem

moderator (5) a component of a light water reactor that slows the neutrons emitted during the fission reaction so that they are more easily absorbed by other nuclei

N

natural gas (3) a combustible gaseous fossil fuel formed beneath the earth's surface

nitrogen oxide (8) any of several compounds formed by the combination of nitrogen, oxygen, and water; a coal combustion by-product

nuclear energy (1) energy released by nuclear fission or fusion

nuclear fission (5) the splitting of the nucleus of a heavy atom, such as uranium, into parts; fission is accompanied by the release of much energy

nuclear fuel cycle (5) the processes of preparing elements for use in reactor operation, using the elements in operation, recovering radioactive by-products from spent fuel, reprocessing remaining fissionable material into new elements, and disposing of radioactive waste

nuclear fusion (6) the combining of the nuclei of two atoms; fusion is accompanied by the release of a great deal of energy

nuclear power (3) electricity generated by nuclear fission

O

ocean-thermal conversion (6) electricity generation by means of exploiting the temperature difference of the water at the surface of the ocean and that far below the surface

oil (3) a liquid fossil fuel formed by animal and vegetable matter beneath the earth's surface; petroleum

outage (9) a failure or interruption in electricity output

overcapacity (9) the state of having more electricity generating units than are necessary to meet demand

P

particulate (8) a tiny particle of solid or liquid airborne material

peak load (4) maximum electricity demand

photovoltaic cell (6) a cell that generates electricity directly from the sun's light energy

photovoltaic conversion (6) the process of converting the sun's light energy directly into electric energy

plasma (6) matter that is heated beyond the gaseous state to a completely ionized mix of electrically charged particles

plutonium (5, 6) a radioactive metallic element used as fuel in liquid metal fast breeder reactors; a product formed by the absorption of free neutrons by the nuclei of uranium-238 atoms during the fission process in light water reactors

pollution (8) the action of causing something to become physically unclean or impure; the condition that results

potential energy (1) the energy stored in an object by virtue of its position

power tower (6) a central tower that receives the sun's light from reflecting mirrors and converts its heat to a working fluid that drives a turbine; a name applied to the central receiver solar-thermal conversion technology

pressurized water reactor (5) a light water reactor in which the heat from the fuel rods is carried by the cooling water, which is pressurized to prevent it from boiling, to a secondary loop of unpressurized water, which is thereby heated to steam to drive a turbine

price (2, 3) the amount of money given or set as consideration for the sale of an item or service; the cost at which something is obtained

public policy (9) a course of action or plan to determine and guide present and future decisions affecting all people in a community or nation

pumped-hydro storage (7) a storage system in which off-peak electricity is used to pump water above a dam; when later released through the dam, the water drives a turbine to generate electricity

pyrolysis (6) a biomass conversion process in which vegetation is heated in the absence of oxygen

Q

quad (2) an informal unit of energy measurement; short for one quadrillion Btu

R

radiant energy (1); **radiation** (8) energy transmitted by waves or particles

rate base (4) the total value of a utility's generation, transmission, and distribution equipment

reactor year (8) the operation of one nuclear reactor for one year

rem (8) a unit of ionizing radiation; acronym for roentgen equivalent man

renewable resource (3) a power source that is continuously or cyclically renewed by nature

reserve (4) an excess of capacity beyond actual load

reserve margin (4) the percentage of generating capacity available in excess of peak load

resource (3) a source of supply; an available means

right-of-way (7) the corridor of land taken up by a transmission system

S

scrubbing (8) a process in which coal combustion gases are passed through tanks containing a material, such as lime or limestone, that captures and neutralizes sulfur dioxide

semiconductor (6) a material, such as silicon, whose electrical conductivity can be greatly increased by exposing it to heat, light, or voltage

sludge (8) a thick, pasty waste product of coal combustion

slurry (6) a suspension of coal particles in a liquid solvent

solar power (6) the conversion of sunlight into heat or electricity

solar-thermal conversion (6) a process by which the heat energy of the sun is converted to electric energy

step-down transformer (7) a device for changing the small current at high voltage that is transmitted from power plant to local substation into larger current at lower voltage for consumer use

storage (6, 7) the process of keeping electricity for future use

substation (7) a facility for the transfer of electricity from generator to consumer by reducing voltages to values appropriate for distribution

sulfur oxide (8) any of several compounds formed by the combination of sulfur and oxygen; a coal combustion by-product

superconductivity (4, 7) the property that allows certain metals and alloys, when cold enough, to exhibit a loss-free flow of electric current

supply (3, 9) the quantities of goods or services offered for sale at a particular time or price; the process of filling a need or want

T

technology (1, 3, 9) a technical method of achieving a practical purpose; the totality of the means used to provide objects necessary for human sustenance and comfort

thermal pollution (8) the discharge of heat, especially the discharge of warm water into rivers and streams, which raises the net temperature of the aquatic system

thermal storage (7) storage of energy as heat

thermophotovoltaic converter (6) a metallic radiating element that, when heated by sunlight, radiates infrared light to which solar cells are most responsive; an efficient solar cell variation

tidal energy conversion (6) the process of converting the kinetic energy of tides to electric energy

tokamak (6) a nuclear fusion technology in which a plasma is confined in a doughnut-shaped vacuum chamber by means of a system of electromagnets

topping cycle (5) the higher-temperature cycle in a combined-cycle system

torus (6) the doughnut-shaped vacuum chamber used in a tokamak fusion system

transformer (7) a device for converting variations of current and voltage in a primary circuit to variations of current and voltage in a secondary circuit by means of linking magnetic fields and transferring electric energy

transmission (7) the transport of bulk electricity from generator to substation

turbine (4) a device consisting of blades attached to a shaft that converts the kinetic energy of a working fluid (liquid or gas) to mechanical energy

U

uncertainty (9) the quality or state of being uncertain, of not knowing beyond doubt

undercapacity (9) the state of not having enough electricity generating capacity to meet electricity demand

uranium (3, 5) a radioactive metallic element used to produce electricity by means of nuclear fission

V

visibility (8) the degree of clearness of the atmosphere

voltage (1) potential electric energy

W

watt (2) a measure of the rate of electricity generation or consumption

watthour (2) the energy that will be expended by using 1 watt of electricity for one hour

wave energy conversion (6) a technology that uses the kinetic energy of wind to drive turbines that need no working fluid

working fluid (4) a liquid or gas that provides kinetic energy to a turbine for conversion to mechanical energy

SUGGESTED READING

Alder, Irving. *Electricity in Your Life*. New York: John Day, 1965.

Asimov, Isaac. *How Did We Find Out About Nuclear Power?* New York: Walker, 1976.

Asimov, Isaac. "The Nightmare of Life Without Fuel." *Time*, April.25, 1977.

Behrman, Daniel. *Solar Energy: The Awakening Science*. Boston: Little, Brown, 1976.

Burke, James. *Connections*. Boston: Little, Brown, 1978.

Carr, Donald E. *Energy and the Earth Machine*. New York: W. W. Norton, 1976.

Clark, Wilson. *Energy for Survival*. Garden City, N.Y.: Doubleday, 1974.

Comar, Cyril L. "Plutonium: Facts and Inferences." *EPRI Journal*, November 1976, pp. 20–24.

"Creating the Electric Age." *EPRI Journal*, March 1979.

Crowley, Maureen, ed. *Energy Sources of Print and Nonprint Materials*. New York: Neal-Schuman Publishers, 1980.

Dunsheath, Percy. *Giants of Electricity*. New York: Thomas Y. Crowell, 1967.

"Earth's Renewable Resources." *EPRI Journal*, December 1981.

"Energy—A Special Report." *National Geographic*, February 1981.

Energy in Transition 1985–2010: Final Report of the Committee on Nuclear and Alternative Energy Systems, National Research Council. For the National Academy of Sciences. San Francisco: W. H. Freeman, 1980.

The Farallones Institute. *Integral Urban House: Self-Reliant Living in the City*. San Francisco: Sierra Club Books, 1979.

Gallant, Roy A. *Exploring the Weather*. Garden City, N.Y.: Doubleday, 1969.

Halacy, D. S., Jr. *Earth, Water, Wind and Sun: Our Energy Alternatives*. New York: Harper & Row, 1977.

Hamilton, Roger. "Can We Harness the Wind?" *National Geographic*, December 1975, pp. 812–829.

Hopkinson, Jenny. "Turning to the Sun for Power." *EPRI Journal*, June 1979, pp. 18–21.

Kenton, John. "Capturing a Star: Controlled Fusion Power." *EPRI Journal*, December 1977, pp. 6–13. See also editorial by Richard E. Balzhiser, pp. 2–3.

Kenton, John. "The Birth and Early History of Nuclear Power." *EPRI Journal*, July/August 1978, pp. 6–15. See also editorial by Milton Levenson, pp. 2–3.

Kiefer, Irene. *Energy for America*. New York: Atheneum, 1979.

Laliberte, Margaret. "Solar Update." *EPRI Journal*, June 1981, pp. 12–15.

Lawrence, Christine. "DOE Explores Solar Photovoltaics." *EPRI Journal*, September 1980, pp. 27–28.

Lihach, Nadine. "More Coal Per Ton." *EPRI Journal*, June 1979, pp. 6–13. See also editorial by Kurt Yeager, pp. 2–3.

Lihach, Nadine. "Fluidized-Bed Combustion." *EPRI Journal*, December 1979, pp. 6–13. See also editorial by Shelton Ehrlich, pp. 2–3.

Lihach, Nadine. "Coal Supply: New Strategy for an Old Game." *EPRI Journal*, October 1980, pp. 6–12. See also editorial by René Malès, pp. 2–3.

Mawson, Colin. *The Story of Radioactivity*. Englewood Cliffs, N.J.: Prentice-Hall, 1969.

Metzger, Norman. *Energy: The Continuing Crisis*. New York: Thomas Y. Crowell, 1977.

Millard, Reed. *Solar Energy for Tomorrow's World*. New York: Julian Messner, 1980.

Nesbit, William. "Going With the Wind." *EPRI Journal*, March 1980, pp. 6–17. See also editorial by John E. Cummings, pp. 2–3.

"Nuclear Safety After TMI." *EPRI Journal,* June 1980.

Pringle, Laurence. *Energy: Power for People.* New York: Macmillan, 1975.

Pringle, Laurence. *Nuclear Power.* New York: Macmillan, 1979.

Rothman, Milton A. *Energy and the Future.* New York: Franklin Watts, 1975.

Sawhill, John C., ed. *Energy Conservation and Public Policy.* Englewood Cliffs, N.J.: Prentice-Hall, 1979.

Solar Energy. Palo Alto, Calif.: Electric Power Research Institute, 1981.

"Solar Technology Today: Special Report." *EPRI Journal,* March 1978.

Stevens, Charles B. "The Tokamak: Bringing the Star Power of Fusion Down to Earth." *The Young Scientist,* December 1980, pp. 6–11.

Teller, Edward. *Energy From Heaven and Earth.* San Francisco: W. H. Freeman, 1979.

Terra, Stan. "CO_2 and Spaceship Earth." *EPRI Journal,* July/August 1978, pp. 22–27.

Thomsen, Dietrick E. "Bumps in the Torus." *Science News,* January 10, 1981, p. 27.

Watts, Alan. *Instant Wind Forecasting.* New York: Dodd, Mead, 1975.

Weaver, Kenneth F. "The Search for Tomorrow's Power," *National Geographic,* November 1972, pp. 650–681.

Weaver, Kenneth F. "The Promise and Peril of Nuclear Energy." *National Geographic,* April 1979, pp. 459–493.

Weiss, Harvey. *Motors and Engines and How They Work.* New York: Thomas Y. Crowell, 1969.

Wellman, William R. *Elementary Electricity.* New York: Van Nostrand Reinhold, 1971.

Whitaker, Ralph. "Coal Gasification for Electric Utilities." *EPRI Journal,* April 1979, pp. 6–13. See also editorial by Seymour Alpert, pp. 2–3.

Whitaker, Ralph. "Reading the Composition of Coal." *EPRI Journal,* July/August 1980, pp. 6–11. See also editorial by George Preston, pp. 2–3.

Wilhelm, John. "Solar Energy: The Ultimate Power House." *National Geographic,* March 1976, pp. 381–397.

Wilson, Mitchell. *Energy.* Chicago: Time-Life Books, Life Science Library, 1963.

"Wind Power." *Energy Researcher.* Palo Alto, Calif.: Electric Power Research Institute, 1981.

Wolfe, Ralph. *Home Energy for the Eighties.* Charlotte, Vt.: Garden Way Assoc., 1979.

Young, Gordon. "Will Coal Be Tomorrow's Black Gold?" *National Geographic,* August 1975, pp. 234–259.

OTHER INFORMATION SOURCES

American Gas Association
1515 Wilson Boulevard
Arlington, Virginia 22209

American Museum of Science and Energy
Science Education Resource Center
Oak Ridge Associated Universities
P.O. Box 117
Oak Ridge, Tennessee 37830

American Nuclear Society
555 North Kensington Avenue
LaGrange Park, Illinois 60525

American Petroleum Institute
2101 L Street N.W.
Washington, D.C. 20037

American Public Power Association
2600 Virginia Avenue N.W.
Washington, D.C. 20037

Americans for Energy Independence
1250 Connecticut Avenue N.W.
Washington, D.C. 20036

Atomic Industrial Forum, Inc.
7101 Wisconsin Avenue
Washington, D.C. 20014

Center for Renewable Resources
1001 Connecticut Avenue N.W.
Washington, D.C. 20036

Department of Energy
National Energy Information Center
1F-048, 1000 Independence Avenue S.W.
Washington, D.C. 20585

Edison Electric Institute
1111 19th Street N.W.
Washington, D.C. 20036

Gas Research Institute
8600 West Bryn Mawr Avenue
Chicago, Illinois 60631

Institute of Gas Technology
3424 S. State Street
Chicago, Illinois 60616

National Coal Association
1130 17th Street N.W.
Washington, D.C. 20036

National Rural Electric Cooperative Association
1800 Massachusetts Avenue N.W.
Washington, D.C. 20036

National Solar Heating and Cooling Information Center
Box 1607
Rockville, Maryland 20850

Office of Consumer Affairs
621 Reporters Building
Washington, D.C. 20201

Office of Technology Assessment
Energy Program
U.S. Congress
Washington, D.C. 20510

Solar Energy Research Institute
1617 Cole Boulevard
Golden, Colorado 80401

Thomas Alva Edison Foundation
143 Cambridge Office Plaza
18280 West Ten Mile Road
Southfield, Michigan 48075

U.S. Environmental Protection Agency
401 M Street S.W.
Washington, D.C. 20460

INDEX

Pollution
    air, 77–81
    from coal combustion, 77–81
    cost of controlling, 87–88
    legislation for, 88
    water, 82
Potential energy, 1
Power tower concept. *See* Central
    receiver system.
Pressurized water reactor (PWR), 33
Price
    and energy demand, 8–9
    and resource use, 13
Proton, 2
Public policy, 90
Pumped-hydro storage, 63, 65
Pyrolysis, 50

Quad, 7

Radiant energy, 2
Radiation, 83–87
Rate base, 24
Reactor year, 85
Reliability, generator, 23
Rem, 83
Renewable resource, 13, 16–17, 36–39,
    44–61. *See also names of individual resources.*
Reserve, 23
Reserve margin, 23
Resource use, 13–14
    future, 17
Right-of-way, 70

Safety, of nuclear plants, 35
Sawdust principle, 20–21
Scrubbing, 79

Semiconductor, 46–47
Shale, oil, 14
Silicon, 46, 47
Sludge, 81
Slurry, 41
Solar power, 44–47
Solar-thermal conversion, 45–46
Solid waste
    management of, 81
    municipal, 50–51
Solvent-refined coal (SRC), 41–42
Steam engine, 3
Steam turbine, 20, 29–31
Storage
    battery, 63–65
    chemical conversion, 65, 67
    compressed-air, 65, 67
    in flywheel, 67–68
    for load management, 64
    pumped-hydro, 63, 65
    of solar power, 45–46
    in superconducting magnet, 67
    thermal, 65
    of wind power, 48–49
Substation, 72
Sulfur oxide, 77–78
Superconducting generation, 21
Superconducting magnet, 67
Superconductivity, 21, 67
Supply. *See* Energy supply.

Technology
    development of, 60–61
    effect of on generation planning, 26
    effect of on resource use, 14
    and energy decision making, 90
    growth of, 3–5
    modern, 4
Thermal pollution, 82

Thermal storage, 65
Thermophotovoltaic converter (TPV), 47
Tidal energy conversion, 50
Time-of-use rate, 24
Tokamak, 59
Topping cycle, 31
Torus, 59
Tower, for transmission and distribution
    system, 70
Transformer, 72
Transmission system, 68–76
Turbine, 19–20
    in combined-cycle, 30–31

Ultrahigh voltage, 70
Undercapacity, and future electricity
    use, 91
Underground transmission, 73–74
Uranium, 16, 32–33, 56–57
Utility costs, and decision making, 90
Utility operation, 22–27

Voltage, 3

Waste disposal
    coal combustion, 81
    municipal, 50–51
    nuclear, 85–87
Water quality, 82
Water turbine, 20
Watt, 7
Watthour, 7
Wave energy conversion, 49
Wind power, 47–49
Wind turbine, 19, 47–48
Working fluid, 20